SCIENCE, TECHNOLOGY AND DEVELOPMENT SINCE INDEPENDENCE

SCIENCE, TECHNOLOGY AND DEVELOPMENT SINCE INDEPENDENCE

S. JAFAR MAHMUD

ANMOL PUBLICATIONS PVT. LTD.
NEW DELHI-110002 (INDIA)

ANMOL PUBLICATIONS PVT. LTD.
4374/4B, Ansari Road, Daryaganj,
New Delhi-110 002

Science, Technology and Development Since Independence

First Edition 1997

ISBN 81-7488-793-8

PRINTED IN INDIA

Published by J.L. Kumar for Anmol Publications Pvt. Ltd., New Delhi-110 002 and Printed at Mehra Offset Press, Delhi.

Dedicated

to

the memory

of

Pandit Jawahar Lal Nehru

the Architect of Modern India.

CONTENTS

FOREWORD

Shri S. Jafar Mahmud has rendered a useful service to lay-understanding of India's achievements in the field of Science and Technology through his book. The general public is aware of the significant progress achieved in the realm of Science and Technology, but this awareness is limited to an appreciation of the more visible aspects of this progress. A great deal has, however, been done which is not conspicuous, but nonetheless, is very important. This book seeks to give a connected account of the multifaceted progress achieved. It describes, of course, our accomplishment in the realm of Atomic Energy, Space Science and Agriculture. It also fills in the details of the work of the Indian scientists and technologies in the realms of Biotechnology, Defence, Lasers, Electronics and Instrumentation, Telecommunications and Non-Conventional Energy. A significant achievement of the book is to invest the narration with a sense of robust optimism and confidence in our future.

It describes the evolution of what Nehru wished for, namely, the "Scientific Temper" in the country. I hope this book will inspire confidence in the decision-makers in Government and the entrepreneurs in industry of the capabilities of Indian Science.

The author rightly conclude that given the will to pursue Research and Development work in various fields, India can achieve a self-reliant economy. This optimistic note is certainly justified when one looks back on the progress we have made in more sophisticated areas like space technology, ocean and explorations and Antarctica expeditions.

It is befitting that at the time of 50th anniversary of Independence that this book will be widely available.

(A. Rahman)
Founder Director
National Institute of Science Technology
and Development Studies (NISTADS)
&
Former Adviser (Planning),
Council of Scientific & Industrial
Research (CSIR)

PREFACE

The importance attached to technological innovation in the industrialised world will be obvious from the following statement of Sir Francis Bacon:

"It would be an unsound fancy and self-contradictory to expect that things which have never yet been done can be done except by methods which have never been tried."

Jawahar Lal Nehru often referred to the paradox of a rich country inhabited by poor and starving people and asked: "why are we unable to convert our natural endowments into wealth meaningful to our people?"

While naming our first swimming pool reactor, "Apsara", Jawahar Lal Nehru reminded the audience that the Elephanta Caves on the other side of Trombay represent the spiritual roots of India and wanted that the power of science and the power of spirituality should become blended in the service of humankind.

Indira Gandhi portrayed the same idea in a different manner when she released a special stamp in 1968 to commemorate the beginning of the wheat revolution in the country. She said: "God was working alone in our fields for centuries. Men and God started working together with the evolution of agriculture over 10,000 years ago. Now, God, men and scientists have combined to produce two ears of wheat where only one grew before."

Thus, our political leaders have continuously underlined the fact that science and technology are the most important components of the dividing wall between poverty and prosperity. This condition has been reflected in the enormous support extended to scientists and technologists during the last 50 years.

The aim of this book is to enable the reader to understand and appreciate the latest Indian technological advances and to provide a clear and thorough survey of the numerous ways in which technology has shaped and is reshaping the country we live in. Recent advances in such fields as nuclear energy, biotechnology, health, lasers, electronics and instrumentation, oceanography, solar energy, agriculture, mining, rural technology and environment management and machine tools and robotics are described. New man-made materials, military and space technology, advances in the field of communication and aeronautics — all these are also covered. An effort has been made to study the Indian developments in a historical perspective and to provide a view of the contributions of Indian science and technology to the world of science during ancient and colonial periods.

As a result of these advances, a subsistence agriculture has been transformed into commercial agriculture, though gaps in the transfer of available knowledge to the fields are apparent. On the health front, a number of serious epidemics like plague and small pox, which used to take a heavy toll of population in the pre-Independence days, have been eradicated. Although India has embarked on developing immunological approaches to contraception, communicable diseases like malaria, leprosy, filariasis and diarrhoea are still on the research agenda.

In the field of nuclear energy, capabilities have been established which span the entire nuclear cycle, namely, exploration, mining, extraction, purification and conversion of nuclear materials, production of fuel elements for reactors, design and construction of power reactors, reprocessing of spent fuel and waste management. Capabilities have also been established in designing satellites and satellite launch vehicles, in providing telecommunications and remote-sensing facilities, and in dovetailing space research programmes to socio-economic problems.

I am deeply beholden to Mrs. Sonia Gandhi for motivation and encouragement to me to write.

I am also thankful to Professor S.Z. Qasim, Professor Satish Chandra, Professor S.M.A. Kazimi, Dr. Kishore Kumar,

Mr. S. Rizwanul Hasan and Ramesh Bhalla for their useful advice. I would also like to thank several people who helped in completion of this book. Mr. J.L. Kumar my Publisher who saw the potential, Ms. Sonali Bhagwati, S.A. Hasan, Imran A. Kazimi, Dr. J.P. Joshi, Mrs. Zahra Hasan, Mohd. Ali Ansari, Vikram Kumar, Dr. C.S. Nanda, S. Azam Zaidi, R.P. Aggarwal, Aniq Hussain, S. Fazle Ali, S.S. Abidi, J.K. Gosain, R.S. Arora, Lalit Bhatia, Ms. Shiba K. Abidi, Ms. Erum Abidi, Shyam Bhatia and S.Khursheed Musanna for their invaluable help in writing this book. I am also thankful to Mr. H.S. Negi and Mr. K.D. Joshi for invaluable help in production of this book.

I am also very thankful to Professor A. Rahman for writing the Foreword.

Thanks are also due to the Library Staff of Nehru Memorial Library, Jawaharlal Nehru University Library, Publication Division Library, CSIR Library and INSDOC Library.

A word of thanks to A. David and Saroj C. David for painstaking secretarial assistance.

S. Jafar Mahmud

Mr. Kulwant, Hemant and Ramesh Singh for their technical aid [illegible]. I would like to thank several people who helped in the preparation of this book: Mr. [illegible] Kumar, [illegible] Vineet [illegible] Prasad [illegible] Agarwal [illegible] for their invaluable help in preparing this book. I am also thankful to [illegible] and Mr. [illegible] for invaluable help in production of the book.

I am extremely thankful to Professor A. [illegible] Salman for writing the Foreword.

Thanks are also due to the Library Staff of [illegible] Nehru Memorial Library, [illegible] Library, CSIR Library and NISCAIR Library.

A word of thanks to [illegible] Sanjay [illegible] David for their [illegible] assistance.

[illegible]

1

HISTORY

India's most spectacular frontier is not only its great cultural and spiritual heritage but also the endless vision of science and technology which has been an integral part of the Indian tradition since time immemorial. While the people of Indus Valley Civilization made use of the wheel and possibly the plough, smelted and forged metal, and developed elaborate fire and flood controls for their farms and villages, today's scientists send rockets hurtling into the same skies mapped more than a thousand years ago by the first astronomers.

The first technological revolution took place in the country between 4000 BC and 3000 BC when India along with Egypt and West Asia emerged from the mists of antiquity. During this period, India was both technically and socially well advanced. Archaeological excavations have shown that there were then thriving cities and centres of commerce. The towns were well planned, the buildings were constructed of standardized fired bricks, and drainage systems were installed. Ports such as Lothal were developed as export centres for early manufactured products of smelted copper and bronze; polychromic paints were used to decorate pottery with a variety of animal and bird designs.

It was not until around 800 BC that the smelting of iron began and then, as today, it was the result of a technological spin-off from the primitive arms industry that existed at the time. In order to improve early weaponry, metallurgical experiments were performed which later led to the use of iron in agriculture and for other crafts. The proficiency of the ironsmiths still stands today in the form of the 24 ft. 3 in. - high six-tonne iron pillar in Mehrauli, Delhi. It is made of wrought iron with an iron content of 99.72 per cent and appears to have been protected against rusting by the application of a thin coating of manganese oxide. A larger pillar, about 42 ft. high, of a later period about 1200 AD now lies broken in several pieces at Char.

The Vedic Age marked a new era of intellectual inquiry and technological endeavour. Religion played an important role in the field of scientific achievements. Ancient mathematical works such as the Sulva-Sutras utilised geometry for designing and constructing altars. Mathematics was an important field of knowledge, and the ancient Indians made valuable contributions to it. Most historians agree that the use of zero originated in India and spread to other cultures. Indians also invented the so-called Arabic numerals (called 'Hindsa' by the Arabs themselves) the knowledge of which reached the West through the Arabs. Mathematicians like Aryabhata and Bhaskara I, Brahmagupta, Mahavira. Aryabhata II and Shrihari used and developed most of the mathematical formulae that we know today. As early as the 5th century AD, Aryabhata I gave the approximate value of pi as 3.1416 — a value which is used to this day. Bhaskara II is well known for his work in algebra and his Siddhantastromani. Indeed, the trails blazed in algebra, trigonometry and geometry ante-dated similar developments in Europe by several centuries.

Astronomy, essential for religious as well as practical purposes, was another field of inquiry which achieved remarkable heights in the ancient times. Aryabhata propounded that the earth rotated about its own axis and calculated the sidereal period of earth's rotation with fair accuracy. Many later scientific works owe their origins to the Panchaskidhanta, of which the Suryasiddhanta greatly influenced astronomical research in India. In the medieval period with the advent of Islamic influence, instruments such as the astrolable, quadrants and armillary spheres came to the used in astronomical research. Raja Sawai Jai Singh II of Jaipur got built in the early eighteenth century, five observatories at Ujjain, Varanasi, Mathura, Jaipur and Delhi, of which the last two are intact to the present. He also produced an elaborate set of astronomical tables, the Zij Mohammad Shahi based on extensive research and astronomical knowledge.

Medicine was yet another field for original research and the ancient Indians made notable advances in it. The Atharvaveda is perhaps the original repository of India's medical knowledge. Study was made of symptoms and causes of diseases, and curative means were researched. Herbs, fruits, flowers and minerals were studied and experimented upon to evolve medical cures. Susruta and Caraka Samhitas, the two great classics of Ayurveda (the science of life), give a clear picture of the medical and surgical practices in use more than 2000 years ago in India. Several surgical instruments, including scalpels, catheters, syringes and forceps, were developed by the early surgeons who could conduct operations for cataracts and laparotomy. Surgical knowledge spread from India to the Arabs, Greeks and Egyptians, Persian influence gave rise to Unani medicine in India in the thirteenth-fourteenth centuries. Plant genetics was also a field of research.

Allied to the medical field was the development of chemistry, a natural fallout of the research in drugs. Chemical knowledge was put to good use in the technological processes of dyeing, production of paper, perfume, and sugar. Experiments were also conducted in the use of new minerals, ores and alloys. Coating copper vessels with tin, creating a new alloy ware, 'bidari', and extracting and using zinc were some of the new developments which were known in India many years before they were used in Europe. From the sixth century BC onwards Indian technical skills were perfected in iron metallurgy and steel, copper, bronze work, ceramics, craftsmanship in precious stones and metals and cosmetics. Several centres for iron forging and copper smelting existed. Construction engineers in the 12th century were using iron girders and beams on a large scale. What is more, they were 99 per cent iron and produced in the same manner as the famous Delhi iron pillar, more than 1500 years old. This pillar at Mehrauli appears to have been made rust-proof by the application of a thin coat of manganese oxide. Unfortunately no record of the forging technique is available.

The ancient Indians developed a variety of technological skills and technical equipment which survive, with slight modifications, to this day. They were able to devise suitable equipment for and methods of agriculture. They were familiar with the growing methods of various crops, treatment of seeds, preparation of soil, crop rotation, irrigation methods. Dikes were built to create reservoirs and irrigation canals were used. Climatic conditions prompted the people to devise means of food preservation in the form of chutneys, murabbas, etc. Ghee was extracted from butter.

In the field of construction the Indus Valley Civilization displayed advanced skills. The Vedic Civilization did not show advanced constructional technique but in time people developed the skill. Cutting and building in stone has always been well developed in India. Stone temples were often constructed on the ground, the stones fitted and marked, then dismantled and reconstructed in the proper setting; the method anticipated the modern technique of reconstructing ruined monuments whose elements can be found and identified. Indian technology developed several very hard cements, e.g. vajralepa. It may be noted that the lacquer technique is said to have originated in India. Constructional engineering reached high levels in the medieval age as reflected in the wide range of monuments, both Hindu and Indo-Islamic.

Hydraulics appear to have been well developed. Many a palace and fort enjoyed the facility of running water, often hot and cold.

It was war, as is so often the case, that led to the development of industrial technology in the 16th and 17th centuries. Cannons and guns of considerable mechanical sophistication began to be made.

Navigation too underwent changes. Indian sailors used a magnetic needle floating on water (shades of the compass?) in the 13th century itself. Significantly many of the European countries had their ships made in India.

However, somewhere along the way, Indians lost or suppressed the inquiring spirit, closed their minds to the free flow of thought: there was stagnation instead of experimentation with new ideas, a preference for a rigid tradition over innovation. Historical and socio-economic factors may have caused this fossilisation: traditional compulsions, instability resulting from frequent political upheavals, lack of a middle class to encourage and support innovative thinkers, a rigid caste-riven social structure, a separation between the thinker and crafsman. Whatever the reasons, it was only with the colonisation of India by England that a new phase in scientific and technological progress began.

The British brought to India contemporary science and technology — what is often termed 'modern' science and technology. However, the educational/research developments in this period were directed to meet the British Government's needs, and not primarily meant for India's socio-economic betterment. But, unwittingly perhaps, these activities promoted indigenous efforts to develop scientific thought.

The foundation of the Asiatic Society in 1784 by Sir William Jones marks the beginning of public interest in scientific research. The Society helped the founding of the Indian Museum of Calcutta in 1866. The Asiatic Society published papers in physics, chemistry, geology and medical sciences and thus played an important role in the advancement of sciences in India.

The Indian Association for the Cultivation of Science, founded in 1876 by Dr. Mahendra Lal Sircar, provided laboratory facilities and became a prominent scientific research centre in the country. One may also recall the Bombay Natural History Society, which was founded in 1883, and Indian Mathematical Society, which was started in 1907 mainly through the efforts of V. Rangaswami Iyer under the name of Analytical Club with its headquarters at Fergusson College, Poona. Calcutta Mathematical Society was established in 1908 with Sir Ashutosh Mukhopadhya as the first President, with the objects of fostering and encouraging the study of mathematics in all its branches, promoting the spirit of original research, and publishing a periodical. The efforts of Prof. P.S. MacMahon of Lucknow and Prof. Simon of Madras led to the formation of the Indian Science Congress Association in 1914. The establishment of these societies played a major role in creating a scientific consciousness, in bringing scientists together and enabling them to make the Government give support to scientific research.

The main scientific activities in the Government sector was largely carried out by the medical and the engineering corps of the army and civil officers interested in science, as a spare time activity. These men, trained in European Institutions and laboratories, left a record of their work and made a mark in various branches of sciences through original contributions. They brought out considerable literature on science and technology, built up a sizeable store of scientific apparatus, chemicals and research tools and founded a few of the important scientific institutions in the country. They created a tradition of dedicated scientific research and the Indians who worked under them carried this forward.

Geologists had been employed since 1818 for survey work. In 1851 the Geological Survey of India was organised, under the efforts of Thomas Oldham, the then Professor of Geology in Dublin. The beginning of geological studies may be traced to the last quarter of 18th century. A series of discoveries of Siwalik fossils, and research on these, were made by H. Falconer and P.T. Cautlay. The Trigonometrical Survey of the Peninsula of India was established in 1800 and was expanded as Great Trigonometrical Survey of India in 1818. The Topographical and Revenue Surveys grouped together under the Surveyor General of India in 1817 and a School of Surveying established in Madras after several transformations, were consolidated with Trigonometrical Survey in 1878 as the Survey of India.

The Botanical Gardens was established in 1788. Dr. William Roxbery was the first to start research on Indian plants in Botanical Gardens. The Botanical Survey of India was established in 1890. Zoological research in India dates back to the appointment of Edward Blyth as the Curator of the Museum of the Asiatic Society in 1841. His successor, John Anderson in 1866 became the first Superintendent of the Indian Museum with the zoological and archaeological collections under his direct charge. In 1910, the Zoological and Anthropological Sections of Indian Museum were converted into the Zoological Survey of India.

Significantly enough, while those areas of science and technology which served British industrial interests were developed early, industrial research and needs of industry were not given any attention by the British, as they wished to keep India as a supplier of raw material to British Industries and market for British manufactures. The requirements of the First World War and the political pressure of the national movement however led to the appointment of the Holland Commission in 1918 for appraising, amongst others, the status of the then existing industrial research facilities, and to make recommendations for its improvement. Nothing much was done till 1935 when the Government established an Industrial Intelligence and Research Bureau with the object of "making a beginning and to lay the foundation on which a research organisation suitable for the needs of the country should later be constructed". An Industrial Research Council was set up to advise on measures for the coordination and development of industrial research.

Once again, the Second World War forced the then Government to establish units to serve the needs of the besieged Britain. The

Board of Scientific and Industrial Research was established to advise the Government on research for development of Indian industries particularly those connected with the war. The Board emphasised the need, and provided the basis for a central organisation to plan research, to bring about effective coordination of the research activities of the country, and to promote the application of research for national development.

In 1942, an Industrial Research Fund was created by the Government for the purposes of fostering industrial development in the country and the Council of Scientific and Industrial Research was constituted as an autonomous body to administer the Fund. The proposals for the establishment of National Physical Laboratory and a National Chemical Laboratory had been accepted, and later plans for other laboratories for food technology, building, road, leather, electro-chemicals and others were formulated. These plans were taken up after independence and research institutes established for these areas, incidentally on the lines of the institutes established in U.K. under the Department of Scientific and Industrial Research.

It was the high incidence of diseases unknown to the west, the cost of their treatment, and their impact on army and administration that necessitated research relating to diseases like cholera, plague, malaria, beri-beri, kala azar, etc. in 1892, the Bacteriological Laboratory at Agra was established with Mr. P.H. Hankin as its head. The spreading of plague in Bombay in 1896 led to the deputation of Mr. W.M. Haffkine to work on this problem. In 1899, Haffkine developed a plague vaccine and established a small laboratory, called Plague Research Laboratory, in Bombay (in 1926, renamed as Haffkine Institute, Bombay). The Pasteur Institute was established at Kasauli in 1900. Three years later, the King Institute was established at Guindy for the manufacture of Calf-lymph and for general bacteriological work. In 1907, another Pasteur Institute was set up at Coonoor. In 1910, Sir Leonard Rogers proposed the establishment of the School of Tropical Medicine in Calcutta. Thus a chain of Institutes with facilities for medical research were established and a cadre of scientific workers in this field created. Sir Harcourt Butler, first Member of the Department of Education, Health and Lands, of the Viceroy's Executive Council and Sir Pardy Lukis, the Director General, Indian Medical Service, worked towards the establishment of an Indian Research Fund Association in 1911, its primary objectives being research, propagation of knowledge and experimental measures generally in connection with the causation, mode of spread and prevention of communicable diseases.

Agriculture research began with the establishment of the Agricultural Research Station and Experimental Farm (later called the Imperial Institute of Agricultural Research) at Pusa, Bihar with the help of a donation made by an American Philanthropist. Mr. Henry Phipps of Chicago. Subsequently, a separate Department of Agriculture was constituted in different provinces. Agricultural Colleges were established at Poona, Kanpur, Nagpur, Layallpur (now in Pakistan), Coimbatore and Sabour. In 1921, agriculture, which had so far been a central subject, was transferred to provinces, which were to deal with policy, administration, coordination of agricultural research and education. A Royal Commission on Agriculture was appointed in 1926 to examine and report on conditions of agriculture and the rural economy of India with particular reference to the measures being taken for promotion of agricultural and veterinary research and education. Its recommendations led to the establishment of the Imperial Council of Agricultural Research in 1920 with the primary object of promoting, guking and coordinating agricultural research and education in India. The Council was also to serve as a link between agricultural institutions in India and foreign countries. A number of Central Commodity Committees dealing with research in particular crops, namely cotton (1921), Jute (1936), sugarcane (1944), tobacco (1945), coconut (1945), and oilseeds (1947), were also set up as semi-autonomous bodies. A Bacteriological Laboratory was established in 1889 in Pune, subsequently transferred to Mukteshwar in 1893. The branch at Izatnagar, Bareilly, was opened in 1913 and the name changed to Imperial Veterinary Institute (now known as Indian Veterinary Research Institute).

Not much research activity was carried on in private institutes. A few institutes were established by scientists or public men, some of them being Indian Institute of Science, Bangalore (1911); the Bose Institute, Calcutta (1917); the Indian Academy of Science, Bangalore (1934) of which the Raman Research Institute is a part; Sheila Dhar Institute of Soil Sciences, Allahabad (1936); the Tata Institute of Fundamental Research, Bombay (1945); Shri Ram Institute for Industrial Research, Delhi (1947). Institutes like the Indian Institute of Science, Bangalore established by the Tatas played a notable role as a centre of research at a time when India possessed few research facilities. Their development was accelerated only after Independence, and they are now centres of higher studies and research in their respective fields.

Though the policy of promotion of science and technology and their use for developing agriculture, health and industry was guided by political considerations, a number of dedicated British scientists found in India unlimited possibilities for contributing to knowledge. Their research and collection of valuable data did much for building a modern scientific base for the country.

Their deliberations, new standards of objectivity-respect for facts as a part of discussions created a new awareness amongst the people, made them realise the importance of science and technology and the possibilities of development through their use. Scientific and technological infrastructure, once created, begins to interact with situations and opportunities and helps in promoting a self-generating scientific and technological tradition. The political leadership of the Independence movement continuously brought pressure on the then Government for greater educational facilities in science and technology, research facilities and creation of industries. As a result, when India became free it had, in contrast to many other colonies, a scientific and technological foundation, which would support the future needs of a newly independent country, provided the political leadership was visionary enough to utilise it.

2

AGRICULTURE

Growth of population and the increasing demand for agricultural products necessitates reclamation of land hitherto held as unproductive. Reduced soil productivity and abandonment of productive soils. Saline and alkaline soils are the result of centuries of neglect and mismanagement. Estimates of salt-affected soils in India vary, but the Central Soil Salinity Research Institute puts it at nearly seven million hectares scattered all over the country.

According to India 1995 till the end of 1993-94, an area of 2.91 million hectare out of a total treatable area of 17.59 million hectare in the catchments of river valley projects has been treated (i.e. 16.50 per cent of the treatable area). In the case of Flood Prone River Scheme till the end of 1993-94, 0.61 million hectare (i.e. nearly 14 per cent) of the treatable area of 4.37 million hectare had been treated.

Alkali soils contain large amounts of carbonates and bicarbonates of sodium, resulting in accumulation of sodium on the surface of soil particles and high pH. The soils have poor drainage and water transmission properties. When cultivated, clods are formed, adversely affecting standing crops. These soils mainly occur in Haryana, Punjab, Uttar Pradesh and Bihar.

Scientists at the Central Soil Salinity Research Institute (CSSRI), Karnal have developed effective, economical and easily acceptable technology for reclamation of alkali soils. Successful reclamation is dependent upon choosing the correct practices for specific problem soils and involves studying the quality and depth of groundwater, calcareousness of soil, and physical condition of subsoil.

The essential components of the technologies for reclamation of alkali soils include: (1) proper water management through provision of irrigation through tubewells, using sprinkler method, light and frequent irrigation; (ii) application of proper amendments such as gypsum, sulphuric acid, sulphur and iron pyrites; (iii) agronomic and cultural practices like land levelling and bunding for uniform water management, effective leaching, increase in dose as well as appropriate application. Increase in dose as well as appropriate application of nitrogenous fertilisers; (iv) choosing the proper crops and varieties during reclamation as crops differ in tolerance to soil deficiencies, salinity, alkalinity; in partially reclaimed soil, special varieties of rice can be grown as kharif and wheat as rabi crop besides sorghum and maize; (v) cultivating fodder grasses and tree species like kikar; an auger hole technique (small holes filled with manure to which a little gypsum is added) for planting trees is effective; (vi) leaching alongwith application of gypsum; (vii) drainage or removal of excess water; this is accomplished when ground-water is used for irrigation, lowering the groundwater table below the critical limits of 2-3 meters.

These technologies in a suitable mix have proved effective in reclaiming alkali soils. Punjab, Haryana and Uttar Pradesh have set up Land Reclamation and Development Corporations to help farmers reclaim alkali lands lying barren for years. Significantly, it has been found that normal production levels can be attained within 3-4 years of reclamation and costs recovered. Reclaimed alkali lands are contributing an additional foodgrains output of more than 2-5 million tonnes annually. Reclamation of alkali soils has environmental consequences; retention and profitable use of runoff water, mitigation of hazards of floods.

Satine soils contain excess of neutral soluble salts including the chlorides and sulphates of sodium, calcium and magnesium. These souls are widespread in the arid and semi-arid regions of Gujarat, Rajasthan, Haryana, Maharashtra, Karnataka and Tamil Nadu. Salinity problems seriously affect many coastal regions due to sea water ingress and shallow saline water tables.

Saline soils have a high water table and the crucial factor in the management of such soils is drainage. The water table may be brought down below the critical limits of 2-3 metres with drainage channels one metre deep and 30 metres with drainage channels one metre deep and 30 metres apart. However, success in reclamation of salt affected soils have been confined to areas which have ground-water of relatively good quality. In areas where groundwater has high salt content, problems of drainage continue to hamper long-term productivity of canal irrigated areas. It is necessary in this context to point out that drainage measures must be integral to any irrigation project; unfortunately in the hurry to spread irrigation to a large area, this vital aspect is ignored. Coping with salinity will depend an in-depth knowledge of the soil-water-plant relationship in a hydrologic unit and the way in which the system responds to interventions for greater productivity.

India, situated in the monsoon belt of south-east Asia, is dependent to a large extent, even now, on rains for its agriculture. Rainfall is unevenly distributed, and in some years the monsoon falls, resulting in drought. It is a measure of the resilience injected into our agriculture by S&T inputs that we are not so severely affected by droughts now as in the past. The geographic situation of the country forces us to learn to cope with inconsistent monsoons. The average annual rainfall we receive is about 370 million hectare-metres of which about 80 million hectare-metres of water seeps into the soil. Half of this remains in the top soil and helps plant growth; the rest goes further down to constitute groundwater. However, it must be noted that saline groundwater is not useful for irrigation, so groundwater must be checked before use. Furthermore, groundwater, according to recent reports, has depeleted alarmingly.

Water being a key input for crop production, its deficiency acts as a bottleneck for the use of other inputs. In India, several irrigation schemes—major medium and minor—have been implemented. By 1984-85, the cumulative irrigation potential had increased to about 675 lakh hectares, and in 1985-90, another 114 lakh hectares had been covered. However, efficiency of most irrigation projects is low, causing a sizeable gap between the created and the utilised potential. Theoretically irrigation should make possible potential. Theoretically irrigation should make possible double and multiple cropping but this is still not the case in much of the irrigated area in India. Water scarcity is an increasingly serious problem.

Obviously the farmers, by and large, lack knowledge of appropriate agronomic practices; supporting facilities necessary for optimum use of irrigation works are not maintained properly; tanks and open wells are sadly neglected. Faulty irrigation practices and absence of proper and adequate drainage facilities have been responsible not only for waste of water but for long-term damage to land through water-logging, salinity and alkalinity, as well as the spread of water-borne diseases.

There was a time when the multipurpose river valley projects were hailed as the harbingers of prosperity. Not any more. These 'temples of Modern India' as Jawaharlal Nehru termed the dams, have no doubt proved an important factor in achieving a successful breakthrough in agriculture and a measure of self-sufficiency in food production. However, their success has been overrated. Serious problems have arisen; heavy siltation of the reservoirs of major dams reduces storage capacity and they become incapable of absorbing heavy floods. A disastrous fallout is the degradation of the soil in the command areas of irrigation projects due to water-logging and soil salinity.

Canal irrigation, again, has contributed to land degradation. Groundwater table rises due to accessions of applied water to groundwater through deep percolation from unlined field channels, distributaries and main canals and restrictive drainage due to construction activities. When the groundwater table rises to within 1-2 metres of the ground surface, it contributes significantly to evaporation from the soil surface, leaving the dissolved salts in the root zone of soils. The salt content of most groundwater in the arid and semi-arid regions is high. This further accentuates the processes leading to the formation of salt affected soils. There are numerous examples of land degradation following the introduction of canal irrigation. The water from the Bhakra canal system was made available in Hissar district in Haryana in 1963. Large areas irrigated by the canal are already salinised. Similar experiences are available in the area of the Tungabhadra irrigation project in Karnataka and in the Nagarjuna Sagar command area. The problems in the Rajasthan Canal command area have acquired serious proportions.

The emphasis on minor irrigation works in recent years is worth noting. Groundwater development is part of this. Minor irrigation schemes, besides being an instant and reliable source of water, also help to control water-logging and salinisation in canal command. States have taken up 'on farm development' works in all the command areas and are introducing warabandhi/intermittent water delivery methods to reduce water application even while increasing farm production.

Sprinkler irrigation is water-efficient and is being introduced in canal irrigated areas of Haryana, Rajasthan and Madhya Pradesh. Drip irrigation is being encouraged in Maharashtra, Karnataka, Andhra Pradesh, Orissa and Tamil Nadu for fruit crops. Fertigation is an innovation of applying fertiliser through drip irrigation to maximise farm productivity with available water. As for surface irrigation, scientists have established that paddy fields do well with a much lower depth of standing water than commonly in practice. Intermittent water delivery, furrow method, microirrigation in well-irrigated areas, delivery of water in pipelines to avoid loss on way — all these can conserve water. The green-house method is another scientific way of growing special crops.

There was a time when you could get certain vegetables in certain seasons only. Not now; these day you get cabbages and potatoes practically the whole year through. By genetic manipulations, scientists have brought about change in the genetic architecture of plants (such as dwarf and bushy) and in the developmental rhythms of plants (for example when the plant flowers) so that they are no longer 'season bound' but 'period fixed.'

These new varieties can be fitted in rotation with other crops, or the simultaneous cultivation of crops can increase production and productivity.

Scientists create ideotypes, or plants with ideal frame and desirable characteristics, chosen after scanning the wide variations in the available germplasm. By incorporating several of the desirable characters in one variety they create a hybrid which may be not only high-yielding but also pest and disease resistant, and of short duration.

There are three stages in the seed production cycle for efficient implementation of the HYV programme. Breeder seed is the primary stage. It is multiplied into foundation seeds to be multiplied finally into certified (equality) seeds which are distributed to all farmers in all regions. The Government set up the National Seeds Corporation (NSC) in 1963 and the State Seed Corporations (CSCs) produce and supply quality seeds. The ICAR organises production of breeder seeds through the breeders and scientists concerned seed technology research was initiated in a coordinated manner in 1979 on receipt of indents from state governments.

The National Seeds Project (Phase III) has been launched with World Bank assistance in March 1990 with the objectives of (i) to assist farmers by ensuring timely and adequate availability of certified / quality seeds of suitable varieties at reasonable prices; (ii) to improve the working efficiency of the national and state level public sector corporations; (iii) to provide facilities for the growth of private sector through adequate institutional finance. Launched throughout India through the Department of Agriculture and Cooperation, ICAR, State department of Agriculture of participating states and NABARD, under the Project, the variety development programme under ICAR and the 'seed certification systems' of state governments are being strengthened.

The policy launched in 1988 is directed towards boosting agricultural production and lays emphasis on (i) import of high quality seeds; (ii) time-bound programme to strengthen/modernise plant quarantine facilities; (iii) effective observance of procedure for quarantine / post-entry quarantine and (iv) incentives to encourage the domestic seed industry to make the country self-reliant in the seed sector.

Biotechnology has a vital role in developing varieties. The World Trade Organisation (WTO) and the intellectual property regime have been established; new varieties will need to be protected either through a patent or a sui generis system. The national institutions will have to compete effectively with multi-nationals in varietal development. Germplasm by itself cannot be considered a resource unless its basic properties are standardised to meet our needs and goals in the plant improvement. In the circumstances, greater investment on the varietal development front would be required from the private sector. The seed industry will play a crucial role in the development of agriculture the world over.

Besides seeds, soil and water, agro-techniques are also important to maximise yield of crops.

Land being limited, it becomes necessary to grow two or more crops a year on the same piece of land to produce enough to meet our needs. Thus multiple cropping practices are to be encouraged. In a relay cropping system, a crop is sown before the harvesting of the preceding crop — this saves water, for the last irrigation for one crop is the presowing irrigation for the other. In mixed cropping, compatible crops are sown in parallel rows with the main crop. Chickpea and mustard can be sown with wheat; pigeonpea, cowpea, mungbean, groundnut, soyabean can be alternated with sorghum, maize, sugarcane, cotton.

Indian farmers have gradually adopted multiple cropping system, but not to the extent desirable.

The major nutrients required by plants are nitrogen, phosphorus and potassium. Plants also need the micronutrients of zinc, boron, iron, molybdenum and cobalt. While their deficiency reduces plant yields, their excess may prove toxic to plants. It becomes necessary to know the nutrient requirements of plants and the nature of soil as far as their supply is concerned. Fertilisers supply plant nutrients in concentrated and readily available form and help higher yield when applied to fertiliser responsive high yielding and hybrid varieties of seeds. A serious attempt to introduce modern crop production technology including the use of chemical fertilisers was made with the implementation of the Intensive Agriculture District Programme (IADP) in 12 selected districts in 1960-61. This was followed by the emergence of high yielding dwarf varieties of wheat and rice and hybrids of maize, jowar and bajra. The introduction of these high-yielding dwarf fertiliser-responsive varieties enabled a real break through in fertiliser consumption in the country. Generalised application of available fertiliser mixture such as the well-known nitrogen, phosphorus, potassium (NPK) combination in the ratio of 120:60:40 can be quite wasteful and counterproductive. The time and depth of application of fertiliser is very important. Unwise use of chemical fertilisers has already resulted in soil micronutrient deficiencies in large areas. Organic manures are essential for maintaining proper soil health. The potential of compost in the country is not fully utilised. Emphasis must be laid on integrated nutrient management (INM) through the combined use of chemical fertilisers, organic manures and biofertilisers.

Organic manures help to improve the physical properties of soil, reduce erosion, increase the moisture holding capacity and, above all these advantages, they are low cost nutrient carriers. But organic measures have their limitations. They are bulky with low nutrient content. The nutrients are released slowly not commensurate with the high and rapid nutrient demand of high-yielding and hybrid varieties of crops.

Use of biofertilisers is not new; they have been used in specific crops like pulses and legume oilseeds for a long time now. Development of biofertilisers received special attention when in the mid-seventies, the prices of petroleum and petroleum products rose highly. These nutrient supply sources have the added advantage that they are renewable and non-pullutant. Biofertilisers are not alternatives to chemical fertilisers but can play a supplementary role in supplying nitrogen to specific crops under specific soil conditions. Nitrogen fixing (non-symbiotic and symbiotic) and phosphate solublising micro-organisms are the main type of biofertilisers that are being used in India. Recently two biofertilisers, namely, Rhizobium cultures and blue green algae have gained popularity amongst the farmers cultivating pulses /legume oilseeds and wet land rice respectively.

Rhizobium which fixes atmospheric nitrogen symbiotically in the roots of legume crops is the best known bio-fertiliser. Considerable research has been done in India to isolate and screen the most efficient strains. It is becoming popular with the cultivators growing pulses and legume oilseeds as it has been observed to give 15 to 30 per cent yield increases. The residual nitrogen effect has been observed in succeeding crops.

Bluegreen algae is another biofertiliser which is gaining popularity in rice growing areas of the southern states, particularly Tamil Nadu, Andhra Pradesh and Karnataka. Blue green algae possess the twin ability of photosynthesis and biological nitrogen fixation. Nitrogen fixed and degraded by algae is taken up by the rice plant. It has, however, not become popular in north-eastern and central India due to unfavourable weather conditions.

Biofertilisers have a limited role in augmenting the nitrogen requirement of various crops. There is, however, scope to enhance its spread, especially in the case of Rhizobium culture for pulses and blue green algae for wet land rice. They cannot replace chemical fertilisers, the major key input for increasing agricultural production.

To promote the use of biofertilisers the Union Government has launched the National Project on Development and use of

Biofertilisers. It consists of one national centre at Ghaziabad and six regional centres at Bangalore, Bhubaneswar, Hissar, Imphal, Jabalpur, and Nagpur for the production, promotion and quality control of biofertilisers. The National Productivity Council has also constituted an award for the best Rhizobium culture producer. The Government also provides grant in aid with institutional agencies, and agricultural universities for the production of Rhizobium culture. Blue Green Algae centres have produced about 350 tonnes annually. These facilities are still inadequate looking to the extent of coverage under pulses and legume oilseeds.

There has been considerable debate on whether the use of chemical fertilisers leads to environmental pollution. Arguments have been advanced that intensive use of chemical fertilisers could result in deterioration of food quality, destruction of natural soil fertility, could cause intestinal cancer, and pollute ground and surface water.

Extensive research studies conducted so far have shown that when fertilisers are used efficiently there is no evidence that they lower the nutritional quality of food. On the contrary in certain circumstances, improvement in the supply of balanced nutrients to the crop can improve the food quality. Similarly when fertilisers are used efficiently in combination with good farm management practices, there is no evidence that they harm the soil. On the contrary, increased, use of chemical fertilisers result in larger crop residues which increase the organic matter content of the soil and hence improve its fertility.

Statistical analysis conducted in advanced countries has shown that there is no correlation between the level of nitrogen consumption in agriculture and incidence of gastro-intestinal cancer. On the contrary, in Western Europe where rates of nitrogen use are among the highest in the world, the incidence of these forms of cancer is diminishing. The pollution of natural waters also depends on how best we can check the nitrogen losses to the minimum through appropriate farming and soil conservation practices. Phosphatic fertilisers have restricted movement and the same is the case with potassic fertilisers.

India with a consumption level of 70 kg/ha is far behind in fertiliser use compared to the developed countries and even China.

Hence, there need not be any apprehension about any negative impact on the environment with the use of chemical fertilisers. Since the consumption of fertilisers is at a low level, greater stress should be laid on efficient and balanced use to avoid any hazards latter on. There are, however, certain pockets under the cultivation of adsali sugarcane in Maharashtra and grapes in Andhra Pradesh, Karnataka and Maharashtra and banana in Tamil Nadu, Maharashtra and Goa where high rates of fertilisers are applied. It would be prudent to check groundwater samples periodically in these areas to test the quality of water. The cadmium content of phosphatic fertilisers used should also be monitored at this stage because this element accumulates in soil and plants and finally travels to the body system of animals and human beings with serious health hazards.

Crop Protection. Crops in the field and in storage are vulnerable to damage from pests. A pest is any organism that causes economic loss in terms of human welfare resources which may be a plant or animal raised for food or fibre. Pests are part of the total ecosystem wherein they survive on species that tend to be abundant and enable to keep ecological diversity. In man-made agriculture, loss from pests is very high. In India alone, annual crop losses are estimated at Rs. 6,000 to 7,000 crores. Of this, losses from weeds are 33 per cent and from diseases 26 per cent, from insects and rodents 26 per cent and birds, nematodes etc., account for the rest. More than 70 per cent pesticides used are insecticides. 12 to 15 per cent fungicides and about 4 to 5 per cent are weedicides. The emphasis in tackling the loss arising from pest needs an orientation on the basis of constraints.

Though the national use of pesticide per hectare is less compared to the developed nations, there are pockets within India where usage is very high. As the pesticides leave residues that persist in the environment and contaminate the food chain, overusage of pesticides is a potential environmental hazard.

No single method of pest control has given a lasting solution. When grown over a large area, resistant varieties break down. Pesticide overuse creates environmental problems, bio-amplification of the residues in human fat and resurgence of pests and results in development of pesticides-tolerant strains of the organism. The time-tested cultural procedures such as growing mixed crops and crop rotation, have been replaced by intensive farming. Changed agronomy, monocropping, cropping sequence, lack of temporal and spatial hurdles have all contributed to pest abundance. It is realised that there is urgent need to evolve a coherent pest management programme that is environmentally friendly and yet able to curtail loss. From out of this concern has emerged a new strategy called the "Integrated Pest Management" (IPM). The IPM is an integrated and planned execution of different pest management strategies and tactics with the sole objective of reducing losses from pests with minimum ecological implications.

Agronomic adjustments help in the avoidance of pests by modifying the macroclimate. For example, flooding the banana field for quite some time, prior to fresh planting, minimises the severity caused by a pest. Addition of organic amendments such as green manuring or neem cake at one tonne / ha of rice crop substantially reduces the severity of sheath blight, and the population of the root feeding nematodies that predispose the plants for infection. Against most of the soil-borne diseases, organic amendment reduces the diseases severity.

Use of high quality certified seeds free from disease is one of the primary means of disease control.

Pest resistant varieties are the core of IPM activity targeted to curtail crop loss. Gene deployment, varietal mosaic, multilines, are the various techniques to increase the life span of a variety.

It must be ensured that crops grown one after another do not have common pests. Pesticides should be applied only when the pest population has reached its threshold level — that is, the level after which they would cause damage. If the threshold level can be predicted, using statistical and computer models timely action can be taken to cut losses. Instead of chemicals, biological organisms that attack the pest can be used. Biocontrol involves employing the natural enemies of crop pests such as parasites and predators, beneficial bacteria and virus in the management of pests. Need-based application of selective chemicals do not affect the non-target organisms that are beneficial to crops. A host of plant-based pesticides includes neem-derivatives, trap crops to isolate pests, pheromones for pest monitoring and management are all included in the IPM strategies. Also included are mechanical methods and use of light traps.

Among the cases of IPM success are the packages for rice, sugarcane, cotton, citrus and a variety of vegetables. The incidence of Heliothis on cotton and gram crops has been controlled. Sugarcane pyrilla, sugarcane borers and rice stem borers have been kept under check.

The revolution in genetic engineering in plant varieties has certainly brought about increased productivity.

Rice production has shown consistent rise over the years : quality too has improved. This is mainly due to the new high-yielding varieties (HYV). The dwarf varieties are not only significantly superior to the tall indices in responding to fertiliser, but they also give higher yield even under low doses of fertilisers. They also have a high harvest index, i.e. ratio of grain to straw. The new varieties made it possible for rice to be cultivated in non-traditional areas. However, these did not suit the traditional north-eastern region as they were susceptible to pests in the humid conditions besides the conditions of lack of optimal use of fertilisers and poor water management. Research is further required to produce suitable strains for this region. Examples of HYV : IR8, Jaya, Heera, VL Dhan 221, CST7-1, Jawahar, TRC 246-C10.

Wheat has shown a remarkable rise in production after the introduction of new plant types such as the dwarf Mexican varieties. The emphasis in breeding wheat varieties in India has been on developing resistance to rusts. It was the spurt in production of wheat that is described by the term 'green revolution'. Some examples of new varieties are MIKS 11. KML 7406, HUW 318 (released for cultivation in non-traditional areas like Nilgiri and Palni hills and resistant to all the rusts). MACS 2496 (which can tolerate high temperatures).

COARSE GRAINS too have shown increased production despite shrinkage in area under cultivation. This is due to the new improved varieties. Some examples — drought resistant SPH 504 (sorghum). MLBI I 104 (pearl millet), GPUK 3, VL 149 (small millets). The new varieties combine short duration with high grain yield and drought resistance.

PULSES are good sources of protein and thus important as a part of the diet in India where the population is largely vegetarian. They are also important for feeding livestock. Being legumes, they maintain soil fertility. Unfortunately, per capita availability of pulses has gone down over the years as they are mainly grown in rainfed areas. Pulses are sensitive to diseases, pests and even environmental hazards such as cold and drought. They thus carry the image of highly risk prone crops. A series of new cultivars are now available which possess resistance to one or more of these hazards at the same time. The traditional varieties of pigeon pea or arhar have too many leaves and branches and few pods: the improved varieties are compact and manageable with more pods and seeds and mature earlier.

A change in the developmental rhythm has made it possible to grow mung and urad in winter and summer and rajmash in the plains. In chickpea or chana it is possible to grow plants which are erect with many branches and pods from base to tip. Examples of new cultivars; BG244, ICC 34 (Gram); LBG17, ADT3 (urad), MIZ 67, CO4 (mung), Rachna (field pea).

Cultivation of pulses, however, requires efficient crop management besides good varieties of plants. Timely sowing, irrigation at critical juncture, adequate doses of phosphate, proper spacing of plants, need-based pest control, are all important; unfortunately, the management technology has not spread among farmers as fast as it should have.

OILSEED production, though on the rise, is not enough to meet the requirement of the population. For very long now the cultivation of oilseed crops has been confined to marginal lands poor in nutrients and water supply. Thus our oilseed types are those capable of surviving in adverse conditions, but their production is low. However, the Technology Mission on Oilseeds (TMO) has improved the situation to a large extent. The TMO worked closely with the ICAR to improve available technology. ICRISAT Collaborated with ICAR to demonstrate high yield capabilities of its varieties of groundnuts — ICGS-11. ICGS-44 etc.

The CSIR network is developing need-based processing technology, improved oil extraction methods, and better oil ghanis.

The strategy should be to fit oilseeds in rotation with other crops in areas where good soil and assured rainfall are available.

SUGARCANE production has shown considerable progress. India has pioneered research in superior varieties and produced several strains including the internationally popular NCO310. Breeding for resistance to red rot has led to varieties such as Co 975 to Co 62399 which increased cane production in problem areas. In order to ensure optimum population and growth of cane, appropriate planting techniques have been developed for different situations. A new approach to evolving a package for integrated pest and disease management in sugarcane is in progress. In this strategy, key pests and diseases in different agro-climatic conditions will be monitored. Precise information on their threshold levels for economic damage in the light of prevalent varieties and environment will be obtained and then varieties and environment will be obtained and then decisions will be taken on the management programme.

Sugarbeet, introduced in India in the 1960s, can be grown in soils with high sodium content, for it removes excessive sodium and improves the texture of such soils.

Cotton is our foremost fibre crop. India is the first country to have developed hybrid cotton. However, the per capita availability of cotton has decreased. Limitations are : it is a rain-fed crop, irrigated cotton, is only 30 per cent in area, pest damage affects hybrid varieties. Also, quality of ginning should be improved.

Jute is a traditional crop, but its demand has been adversely affected by development of other materials for packaging, changes in storage and transportation. However, an environmentally conscious world is now showing renewed interest in jute. The Central Research Institute for Jute and Allied Fibres (CRIJAF). ICAR, played a pivotal role in developing improved production technology for both jute and mesta. High yielding varieties apart, the CRIJAF has also developed a package of practices recommending fertiliser schedules for different situations, foliar fertilisation, row cropping with multi-row seed drill, pest-disease management and improved retting and extraction techniques. Besides, a jute based cropping system with several crop-rotations under both rainfed and irrigated conditions has been advocated for maximisation of farm income. Some examples are : jute-paddy-lentil, jute-arhar, jute-black-gram (under rain-fed); jute-paddy-wheat, jute-paddy-mustard (under irrigation). Recently, the Jute Technologica Research Laboratories (JTRL), ICAR, has developed a technique for application of fungal culture at the post-retting stage for improving the quality of barky jute.

Potato is propagated by seed-tubers. The aphid pest causes much damage through viral disease, Scientists have now identified different periods of the year when different regions of the country are free from aphids. With the development of suitable varieties, all potato growing regions in the country (and not just hilly regions) can grow healthy seed-tubers in aphid-free times. Scientists have also developed the technology of raising the crop from potato seed (true potato seed technology or TPS) and not just cut pieces of seed tubers. This will cut the cost of cultivation of potato besides spreading it as much in the plains as in the hills. It has also eliminated the danger of seed borne diseases.

Horticultural products is general and fruits in particular yield more calories and nutrients required in human diet than the field crops per unit area. Further, the perennial crops, namely, fruit and plantation crops as components of social forestry help maintain the ecological balance and the fight against atmosheric pollution. Horticultural crops, particularly fruits, have great export potential and can earn foreign exchange in sizeable quantities, if the existing resources are tapped to a greater extent.

Considerable progress has been made in horticultural development during the past two decades. Horticulture educational has been strengthened with the creation of horticulture faculties in many State agricultural universities. A separate University of Horticulture and Forestry has been established in Himachal Pradesh. The Horticulture Cell at the ICAR head-quarters has been strengthened with the creation of the post of Deputy Director General of Horticulture who coordinates horticultural research in the entire country. A number of research institutions and national research centres have been set up to conduct studies on various horticultural crops.

Productivity of fruits is very low in India — a mere 9.05 tonnes per hectare. The common contributory factors for the low productivity are : (a) plantations with poor quality planting material; (b) poor orchard efficiency; (c) weak extension linkage between research and farmers; (d) lack of data base for identification of crops with specific locations.

Realising the importance of horticulture and the need for research back-up, the Indian Institute of Horticultural Research

(HHR) was set up at Hessaraghatta near Bangalore in 1968. Regional stations of IIHR have come up at Chethalli / Gonikoppal in Karnataka, Godhra in Gujarat, Ranchi in Bihar and Lucknow in Uttar Pradesh. The regional station at Lucknow in Uttar Pradesh. The regional station at Lucknow was later upgraded to an independent institute and named the Central Institute of Horticulture for Northern Plains. Four national research centres for citrus, banana, temperate fruit and arid fruits were sanctioned during the Seventh Plan.

Research on fruit improvement by breeding, standardisation of production technology, crop protection, post harvest handling and processing of fruits has been intensified during the past two decades. Major emphasis has been given to dryland horticulture, particularly semi-arid fruits.

Breeding in mango aimed at overcoming the problem of alternate bearing and evolving dwarf genotypes for high density planting and increased productivity per unit area has resulted in the evolution of several varieties like 'Mallika' and 'Amarapali', Cultural practices to overcome mango malformation have been standardised.

In banana, breeding of varieties resistant to bunchy top and panama wilt has been initiated. Tissue culture techniques for the mass multiplication of plants have been standardised and commercially produced.

In grape, the optimum time for pruning to avoid diseases and obtaining high yields has been standardised. The optimum dose of nutrients has been worked out for different varieties.

In semi-arid fruits, a breakthrough has been achieved in the production of ber and pomegranate. Marginal lands in semi-arid regions are utilised to produce high levels of crop in ber varieties like 'Umran' and 'Gola'. A good example of this type of revolution could be seen in Ahmednagar district of Maharashtra, where marginal and waste lands have been converted to excellent pomegranate orchards.

Strategies for horticultural development should mainly aim at nutrition security to the growing population and the export of horticultural produce. The daily per capita consumption of fruits in India is 40g as against 120g recommended by the Diet Advisory Committee of the Indian Council of Medical Research. At the estimated production of 26.6 million tonnes of fruits and with a population of 840 million, the per capita availability of fruits should have been 86.8g. The big gap between what should have been the consumption (availability) and the actual level can be attributed to post harvest losses to a great extent. Factors contributing to such losses are (a) not harvesting fruits at the correct stage due to lack of knowledge on maturity standards: (b) improper handling and lack of field cooling facilities; (c) lack of post-harvest treatment with fungicides; (d) lack of refrigerated transport. (e) non-utilisation of inpacking material. SO_2 releasing pads and ethylene absorbents; (f) using improper containers; (g) non-availability of controlled atmospheric storage; (h) monoculture of varieties and crops resulting in glut; (i) limited utilisation in processing. These aspects need to be kept in mind while standardising technology packages to counter post-harvest losses.

Increased productivity can be achieved by growing high yielding varieties, mass multiplication of quality and true-to-type planting material by tissue culture and evolving suitable agro-techniques including high density planting and use of chemicals and growth regulators.

Vegetable production appears quite impressive in that India ranks next only to China in this regard — around 51 million tonnes from a cropped area of 4.5 million hectares. However, per capita availability is around 120-130 g/day against dieticians' recommendation of 300 g/head/day. This is due mainly to the low productivity levels in vegetable crops.

In order to provide the necessary research backup for vegetable production, a number of research institutions were established in the country after Independence. Vegetable research is now carried out at four Central Institutes, one National Research Centre and 26 State agricultural universities. The All India Coordinated Research Programme of the Project Directorate of Vegetable Research offers facilities for multidisciplinary, area-specific research on 23 vegetable crops and provides a national grid for multilocation testing of technologies developed by various institutions.

The development of a large number of improved varieties with wider adaptability and standardisation of their production technologies for various agro-climatic conditions has made it possible to produce vegetables in wider areas and throughout the year, and has improved the prospects of their supply tremendously. A number of disease and pest resistant varieties have been evolved to reduce the use of pesticides. The Directorate of Vegetable Research has provided details on agronomic practices, such as spacing and fertiliser recommendations, for better production.

If despite the development of improved technology (high yielding open pollinated varieties and hybrids and agronomic practices, the productivity achieved by the Indian farmer is rather low, the reasons are :

(i) Inadequate organisational set up. This was quite serious till the Eighties. Now there is a DDG (Hort) in ICAR, a Horticultural Commissioner (NHB), departments of horticulture in a majority of the States and also departments of vegetable crops in agricultural / Horticultural universities.

(ii) No reliable data on area, production and productivity. These are the basic inputs for planning and execution of a programme.

(iii) Inadequate irrigation facilities. These are improving slowly.

(iv) High price and low usage of fertilisers.

(v) No linkage with other rural development programmes.

(vi) Inadequate extension service.

(vii) Lack of organised marketing and transport.

(viii) Inadequate quality seed.

(ix) Lack of prevention of post-harvest losses.

(x) Lack of suitable varieties for processing and export.

(xi) Lack of cargo storage and cargo space for export.

(xii) Inadequate budget allocation.

The technological inputs required are: specific schemes to boost vegetable production: strengthening seed production programmes ; extension work and linkage with other rural development agencies; integrated nutrient, water and pest management ; improved post harvest technology.

Priority areas of research are: breeding for resistance of biotic / abiotic stresses; heterosis breeding; breeding for nutritional quality and processing; growing vegetables under protected environment; use of biotechnology; tackling pesticidal residue.

Floriculture or flower cultivation, though an ancient practice in India has not received much attention for generating employment and even foreign exchange.

Important flower growing States are Karnataka, Tamil Nadu, Maharashtra, West Bengal, Andhra Pradesh and Rajasthan. According to the Agricultural and Processed Food Products Export Development Authority (APEDA) in 1988 there were 24,000 hectares under traditional flowers and 10,000 hectares under modern cut flowers and related products. The floriculture trade turnover is around Rs. 205 crores per annum, the traditional plants account for Rs. 105 crores.

Broadly, floriculture products include bulbs, tubers, tuberous roots, corms, crowns and rhizomes; live plants, cuttings and their slips; cut-flowers and flower buds, fresh, dried, dyed, bleached etc; foliage, branches and other parts of plants — fresh, dried, dyed and bleached.

Despite its long history and the favourable agro-climatic conditions in the country, the floriculture scenario is marked by poor agricultural practices and low quality of products. This is attributed, among other things, to the old and in most cases obsolete varieties being cultivated which are low yielding and poor in quality as compared to the modern cultivars grown in Holland, the U.S., the U.K. and France. Admittedly horticulture did not get its due recognition when the agricultural programmes of the five year plans were drawn up and in the horticulture field, research in floriculture continued to be the most neglected.

Prior to the Second Plan, most of the improvement work in floriculture was done by private nurseries and to some extent by the Botanical Gardens at Lal Bagh (Bangalore) and Sibpur (Howrah) and the National Botanical Garden, Lucknow. Some work was done under the Cess Fund Scheme of the ICAR at several centres. During the Second Plan, the Division of Horticulture was created at the Indian Agricultural Research Institute (IARI) which is claimed to have heralded, the beginning of floriculture research. This effort got a boost in 1968, when the Indian Institute of Horticultural Research was set up at Hessarghata (Bangalore). The launching of the All-India Coordinated Floriculture Improvement Project in 1970-71 provided the much needed thrust to these research efforts.

The two decades of research under this project, carried out through 10 different institutions, has led to the evolution of new varieties and development of suitable agro techniques. The Rose, Gladiolus, Carnation, Chrysanthemum and Orchids were chosen for concentrated research in 1982 as these had good export potential. Four more were added during the Seventh Plan for research works, namely, Antirrhinium, Petunia, Tulips and Daffodils.

The project aimed at building up large collections of germ plasm of different crops and evolving new varieties on the one hand and developing improved agricultural practices on the other. While the research efforts started yielding results, their impact at the field level was not visible. There is no centrally sponsored scheme on floriculture in any State. There are only State sector programmes and these were in Andhra Pradesh, Karnataka, Kerala, Maharashtra, Manipur, North East Region, Punjab, Uttar Pradesh and Jammu and Kashmir. These Seventh Plan schemes were mainly for developing gardens and distribution of planting material. There were however two foreign-aided projects on cut-flowers, one in Karnataka and the other in Jammu and Kashmir.

According to the Eighth Plan Working Group Report, the constraints to development of floriculture included : lack of scientific information on flower growing for cut-flowers; inadequate extension services to ensure transfer of research results to the growers; inadequacy of trained personnel; non-availability of quality seeds and planting material in large quantities; lack of infrastructural facilities such as cold rooms to cool the cutflowers, air conditioned trucks for transport, packaging materials; non-availability of greenhouse materials.

Identified floricultural development areas suggested are Delhi, Pune, Nasik, Bangalore, Thiruvananthapuram, Kalimpong, Calcutta, Srinagar, Solan, Coimbatore-Nilgiris.

The private sector has established a tissue culture laboratory for producing hybrid seeds for flowers.

Animal husbandry has been a significant component of agriculture with an important role in rural economy. Livestock provide essential animal proteins through milk, meat and eggs; they provide wool and skins; even their excrement is used as fuel directly or in biogas plants.

Animal science research is related to (i) improving production and productivity of different livestock species; (ii) reducing losses in production and productivity resulting from mortality and morbidity due to diseases; (iii) reducing losses in handling and processing of livestock products.

While the Indian cow constitutes about 15 per cent of world population. It contributes a more 3.2 per cent of the world's cow milk pool. The cattle is by and large low yielding and non-descript with low levels of nutrition. In the circumstances, the ICAR and the National Dairy Development Board (NDDB) have mounted massive efforts to integrate dairy science research, education, extension and development.

Efforts at cattle improvement have produced new strains — the Karan Swiss and Karan Fries — which yield above 3500 kg of milk per lactation.

In buffalo research, an elite herd with an average lactation period of 305 days and yield of over 3000 kg of milk has been established.

Subsequent to the adoption of frozen semen technology, research was taken up in embryo transfer and associated bio-technologies for developing transgenic animals, i.e. animals whose hereditary DNA has been augmented by the addition of DNA from a source other than parental germplasm using the recombitant DNA techniques. Research involves developing techniques for cloning, sexing, in-vitro maturation, multi-ovulation, cryo-preservation of embryo and genetic engineering. The use of biotechnology in the multiplication of high-yielding germplasm is promising. Realising the value of biotechnological techniques the Union Government has launched a project "Cattle Head Improvement for Increased Productivity, using Embryo Transfer Technology" involving the department of Biotechnology, NDDB, NDRI, NRI and National Institute of Immunology, besides several regional centres. The project's laboratories have produced frozen embryos.

Deficiency of feeds and fodder is a major constraint in the dairy development programme. Non-conventional resources have been explored to provide feeds, such as the by-products or wastes of agriculture, forests or industry, after suitable chemical / microbiological treatment, Some examples are fruit canning industry wastes, distillery wastes, salseed meal, rubberseed cake. Recently coconut pith, kokam cake, kosum cake, mahua seed cake, cocoa pods, maize cobs have been identified as feed resources.

Biotechnology related to DNA recombinant techniques, monoclonal antibodies and hybridomas has emerged as an important area of application in animal health research, particularly for the development of suitable diagnostics and prophylactics.

Advances have been made in indigenous dairy products technology research. Bovine milk has been suitably modified to form an infant formula; a purely vegetarian processed cheese using fungal rennet, a synthetic ghee flavour, and a new ghee making plant are other developments. A unique technique has been standardised to preserve raw milk up to 24 hours at 32°C. Acidophatlus bacterium has been used as coagulating agent to increase shelf life of cheese.

Shortcomings that need to be rectified are : not enough good quality feeds and fodders: lack of an organised animal husbandry extension service; lack of machinery to transfer new technologies to the field; funds allocated for research inadequate.

Many Central research institute of the ICAR are producing vaccines to reduce the incidence of cattle diseases such as rinderpest and foot and mouth. The NDDB has established a disease diagnostic and investigation laboratory to work on vaccines. Research is undertaken to improve productivity of milk per animal by ensuring proper feed and nutrition management. Quality forages and leguminous grasses are being evolved. An animal disease research centre is coming up at Bhopal. With the popularisation of embryo-transfer technology, even barren cows and buffaloes may produce high-yielding progeny.

Poultry. The poultry industry has shown a phenomental growth over the last two decades. With land holdings getting smaller and agricultural operations becoming non-viable, the farmer has to choose alternative methods with limited land and water use. Poultry farming is the best suited animal agriculture which could be favourably adopted. A farmer can grow as many as 1.2 lakh brotlers in six crops or 10,000 layers per annum in just one acre. The poultry industry utilising the genetic benefits of hybrid brotler/egger chicks with the application of biotechnological techniques, has provided a solution for the health hazards of meat eating. The organised growth in poultry had created a new dimension of food safety for poultry meat.

The modern hybrid hen, it is said, evolved from the Red Jungle Fowl of India, but organised poultry farming has become popular only in recent times. With a production of about 26.1 billion eggs in 1994-95, India stands fifth in the world. Indian per capita consumption of eggs is 32. Fundamental and applied research has improved poultry farming. A well-planned effort at indigenisation and acclimatisation has improved poultry farming. A well-planned effort at indigenisation and acclimatisation has improved the productivity of layers as well as brotlers. The poultry sector makes a major contribution to the agricultural sector in the form of organic manure. Poultry is a rich source of protein in our diet. The growing poultry industry is set to benefit from further research and application of biotechnology.

Ducks are reared both for egg and meat. The egger breeds are more prolific layers than the best of chicken breeds. Ducks have a long productive life. They maintain their egg production almost at the same level for three production years in contrast to chicken which lay for up to one and half years only. Moreover, they lay large sized eggs. Ducks are more resistant to diseases than chicken and can be reared with less attention. Ducks consume more feed than chicken but grow faster. Indian Runner and Khaki Campbell are the two important breeds of duck used for egg production.

The Japanese quails need small space and are highly suitable for cage rearing. Therefore, it is possible to rear them even in urban localities where space is a major constraint.

Of late, research and development programmes have been increasingly directed towards the application of biotechnology as well as automation of the poultry, duck and quail industries with larger capital input. However these technological improvements should be aimed at providing sustainability and food security to the people from the economic, social and environmental viewpoints.

Besides the Government, several private entrepreneurs also have significantly contributed to the growth of the industry by providing genetically superior hybrid varieties, vaccines, disease control, improved housing and equipment, quality feed as well as marketing of poultry products. It is envisaged that better designs of poultry houses and automation by using engineering and computer technology would be made available in the near future. Technological advances in poultry production have made poultry meat the best of the meat varieties in the world. A chick rearing shed provided with micro environmental control and automation of water circulation is seen in India with private entrepreneurs.

Sixty per cent of the poultry meat/egg production depends upon the feed. Hence due care has to be taken for the formulation of the right type of feed which should ensure optimum utility. Extensive research has been done in the universities and other R&D stations of feed manufacturers to identify unconventional feed which could safely be incorporated in the poultry feed. In fact, attempts are being made even to recycle some of the farm wastes, slaughter house coproducts and available biomass in a productive poultry feed. Alternative methods to fortify energy value of the feed will be to pellet the poultry feed, especially by steam pelleting. Pellet is the compressed form of feed which helps in partial stabilisation and reduction in volume (bulkiness). The other advantages include suitability of automation, prevention of clogging of feed in the feeders, destruction of micro-organisms, and reduced feed wastage since the fallen feed pellets can be picked and consumed by the birds. Each pellet is a true representative sample of the balanced feed.

In certain tracts such as arid, semi-arid and drought prone areas, crop farming alone cannot provide sufficient income for the material welfare of farmers, due to the high aridity, vagaries of weather and poor productivity of land for crop production. These areas offer excellent scope for the rearing of livestock, especially sheep and goats.

Sheep population in the country was 49 million in 1982 as per the livestock census. There are about 32 breeds of sheep. There is wide variation in type of sheep in different parts of the country.

(i) North temperate region—Jammu & Kashmir, Himachal, hilly areas of Western Uttar Pradesh—sheep produce medium to fine wool; breeds; Gurez, Gaddi, Bhakarwal, Rampur Bushair.

(ii) Northwestern region — Rajasthan, Gujarat, Haryana, Punjab, plains of U.P. — coarse quality carpet wool; breeds; Chokla, Magra, Pugal, Nali, Jalsalmert, Patanwadi, etc.

(iii) Southern peninsular region — Maharashtra, Tamil Nadu, Andhra Pradesh, Karnataka — coarse, hairy fleeces or no wool; breeds; Deccant, Mandya, Nilgiri, Hassan, etc.

(iv) Eastern region — Bihar, West Bengal, Assam — sheep useful for meet; better quality wool from sheep in Arunachal Pradesh.

Sheep and goats breeds of the country in the ecologically fragile zones constitute unique genetic material which must be preserved. These breeds have developed the adaptability traits of low fertility, small size, light weight; low production and high utilization of coarse fibres.

Modern sheep breeding practices based on scientific methods which would lead to qualitative improvement in the genetic stock are often difficult for the sheep breeders who are generally migratory in the northern regions.

With a view to improving the production and productivity of indigenous sheep, cross breeding with exotic sheep of fine wool, dual purpose and mutton varieties was initiated in the pre-Independence period. However, systematic programmes in this field were initiated only after the five year plans were taken up. Since then, sizeable sheep development programmes have been taken up for rapid improvement in the quality and quantity of wool. In Jammu and Kashmir, Himachal Pradesh and the hilly regions of Uttar Pradesh, cross breeding work is being done on a large scale, either through natural service or artificial insemination where such facilities are available. Such work is also in progress in Rajasthan, Gujarat, Haryana, Bihar, Madhya Pradesh and Maharashtra for the production of quality wool. Selective breeding has also been according importance in the case of important carpet wool breeds such as Marwari, Jaisalmeri, Pugal and Magra.

The main thrust of the policy on sheep development is the improvement of wool both in regard to its quality and quantity. Cross breeding of indigenous sheep is resorted to, for bringing about qualitative and quantitative improvement in wool production.

Crossing of indigenous sheep with fine wool exotic sheep like the Australian Merino / Rambouillet, lead to the production of dual purpose breeds. Selective breeding is also practised among important indigenous breeds for medium strong wool and mutton.

It is also necessary to consider measures for improved grazing of sheep. This can perhaps be done in two ways — by bringing additional land under pasture either by converting portions of fast degrading sub-marginal lands presently under cultivation into grasslands or by developing pastures in areas which have remained as wastelands.

The Task Force set up to evaluate the impact of sheep and goat rearing in the ecologically fragile zones was of the opinion that lack of adequate marketing facilities adversely affected the production and productivity of sheep and goat. This brings into focus the importance of organising marketing of sheep and wool products through apex State level Cooperative Wool Federations or Wool Boards / Corporations.

Goat development in the country has suffered on account of two conflicting views about the role of the goat in land use. One School of thought holds that goat is the major cause of denudation of forests / range land causing soil erosion and that its multiplication should be restricted. The other view considers this charge as unrealistic and unscientific; that the goat consumes the scrub left over from the grazing of cattle and is a major supplier of animal protein and other important byproducts. Goats can survive in ecologically harsh regions, where other livestock fail to sustain themselves.

The population of goats increased from 447 million in 1951 to 96 million in 1982. This increase in number has occurred in spite of the fact that no special development programme is being implemented and that nearly 40 per cent of the goat population is slaughtered every year.

The striking feature about the distribution of goats in India is that the activity of goat rearing is sustained in many different kinds of environment, dry, hot, wet or cold, high mountains or low laying plains. The activity is also associated with different farming systems —crop based or animal based, pastoral or sedentary, single animal or mixed herds, small scale or large scale.

Goat is mainly reared for meat, milk, hide and skin. India is one of the four privileged countries which produce raw material for 'Gless kid' leather, which is among the most sought after leather in the international market. In addition to skin, other export items from goat are hair and casings for sausages.

Indigenous breeds of goat being reared in different agroclimatic conditions are efficient converters of scrub material to meat and milk. The Jamnapari, Jhakharana and Parbastar breeds are well known as efficient milk producers. The Black Bengal and White Bengal breeds are favoured for their high quality meat, fecundity and skin. The share of goat milk in the total milk production is about 3.5 per cent.

The per head requirement of nutrients in respect of goats is relatively low. They fit into the resources of small farmers with marginal grazing lands which cannot, on any account, sustain large animals throughout the production cycle. It is not true that goats degrade the environment, but they are probably the ones which survive and adapt to a degraded environment because of their enormous ingenuity to search and find out nutrients from almost anywhere. They are essentially browsers and eat plants which no other animal touches. Degradation of the environment attributable to goats is in fact the result of mismanagement by man.

Cross breeding of low yielding indigenous goat with exotic dairy breeds such as Saanen and Alpine may have to be resorted to for enhancing milk production. While the supply of improved genetic material to goat breeders is essential, development of grazing resources is also important.

The pashmina goat is found in the Ladakh region of Jammu and Kashmir. Selective breeding is proposed to be introduced to improve the quality of pashmina in these areas.

Rabbit rearing in this country is of recent origin, but owing to higher returns on investment it has gained momentum. The rabbit produces one of the finest fibres known as Angora hair, or Angora wool. Rabbits can be classified into two categories—broiler or meat rabbit and wool or fur rabbit. No organised farms have been established to produce animals with high returns. Efforts must be made to expand rabbit rearing so that fur need not be imported.

The pig, long associated with dirt, is in reality an efficient feed converting and fast weight gaining animal, indigenous pigs are poor in performance compared to exotic breeds. Cross-breeding and up-grading programmes with use of exotic pigs such as White York-shire, Landrace, Tamworth and Berkshire have been undertaken to improve economic traits of indigenous swine herds. Pig breeding/ multiplication centres have been set up. Regional bacon factories are operating in Andhra Pradesh, Bihar, Uttar Pradesh, West Bengal, Kerala and Maharashtra.

Fish, whether marine or freshwater, are a good source of protein. During the last three decades, technological developments have helped chalk out new initiatives to increase fish production from all sources — near shore, extended economic zone (EEZ), offshore areas, open seas, river valley systems, village ponds, reser-voirs and coastal and inland saline soils. Fish production has in-creased by more than five times in the last four decades — from 7.5 lakh tonnes in 1950-51 to 46.8 lakh tonnes during the year 1993-94.

There is much scope for increasing catch for with an EEZ of 2.02 million km and a potential catch of 4.5 million tonnes a year. India is just producing about 2 million tonnes in marine fishery. Exploi-tation of marine resources is mostly confined to the coastal belt in the 0-50 metre depth zone.

Remote sensing and satellite imageries, besides vessel based investigations are now being used to locate concentrations of fish. Mechanised fishing boats and vessels have also been introduced for better exploitation of offshore fish resources.

Inland fisheries may be 'capture' or 'culture' types. Capture fisheries refer to the catch from rivers and reservoirs. In culture fisheries the fish are grown in ponds, pools, swamps and are caught when they are of the right size.

While fish production from rivers and estuaries is declining, aquaculture is making rapid strides.

Aquaculture which refers to production of aquatic animals and plants such as fish, shrimp, molluscs and sea weeds in water, can be classified under two major heads — freshwater and coastal. The latter can again be subdivided into two categories, namely, sea farming and brackishwater aquaculture. Seafarming refers to culture of organisms in cages, pens, rafts and longlines in the open coastal waters and bays, whereas brackishwater aquaculture relates to land-based farming systems using salt water from the estuaries and creeks as also from the sea in coastal areas. Coastal aquaculture uses many common species of fish, shrimps and molluscs.

Although India has moved ahead on several fronts in the development of techniques for seafarming of organisms with high production potential, commercial-scale operations are yet to take off. On the other hand, freshwater and brackishwater aquaculture is making substantial progress.

Currently, about 65,000 ha of brackishwater area is under aquaculture in India, of which over 52,000 ha are under traditional farming practices in West Bengal, Kerala, Karnataka and Goa. This is a tide-fed system in the low-lying estuarine areas where the post-larval and juvenile fishes and prawns coming in the canals along with the tide are trapped in the adjoining ponds, allowed to grow for a period feeding on available natural food, and periodically harvested. The average production rate in this system is about 450 kg/ha/year.

The last decade witnessed new developments, albeit slow, in brackishwater aquaculture with the Government coming up with favourable policies, guidelines and incentives and the private sector, including farmers and large enterprises, entering the field. Brackishwater aquaculture came to be identified with the production of a single commodity, the prawn, due to its expanding export market, high unit value and short-term (100-120 days) cropping pattern.

The main requirements of aquaculture are seed, feed and water quality.

Prawn hatchery technology for almost all species used has been developed at research institutes.

Research on prawn/fish nutrition, raw materials, diet formulations, feed preparation, testing and evaluation has received greater emphasis during the last three years at the Central Institute of Brackishwater Aquaculture and in some other laboratories. A number of established animal feed manufacturing companies have diversified into production of prawn feed. Foreign collaboration has also been sought by some.

Water quality management includes all physical, chemical and biological factors, and these influence survival, growth, health and production of fish in the ponds. Water quality is environment related and inputs such as liming for sanitation and correction of hydrogen ion concentration, fertilisation stocking and feeding interact with the pond environment. Indian coastal areas have varied agro-climatic conditions and pond management needs to take all factors into consideration.

The inland saline and alluvial soils which are rendered unfit for agriculture can be effectively used for culture of marine fishes and prawns. This scheme was initiated by CIFE, Bombay and some private farmers of Haryana have been successfully persuaded to take up culturing of marine fish there.

The Central Institute of Freshwater Aquaculture (CIFA), Bhubaneshwar, has made significant advances in breeding and rearing the air-breathing cat-fish, popularly called magur, in a greater part of the country. The biotechnological approach has been adopted and the fish are implanted with hormonal pellets to advance maturity. These are then repeatedly spawned from April to October, in and out of the breeding season. The CIFA has been transferring this technology for the past two years through regular training programmes. Investigations on large scale seed production of the other air-breathing cat fish have been planned. Hypophysation of murrels is well-known and seed production possible. Recently, the CIFA has also bred and reared the freshwater shark, Wallago attu, which is a popular fish in north and north-western India as well as Manipur.

Sewage effluents can be used for culture of the common carp, as evident in West Bengal.

One of the most lucrative and foreign exchange earner enterprise is the production of cultured pearls from freshwater bivalves. The technology developed by the CIFA during the last three years will go a long way in generating employment and bringing about the economic emancipation of rural women. Training programmes for transferring the technology are being planned

Besides, mollusc culture itself will provide a valuable export item.

The introduction of growth hormone gene in fish embryos will revolutionise aquaculture production by enhancing the growth rate, at least, by two to three times. Attempts in this direction are being made at the National Institute of Immunology, the CIFA and the Madurai Kamaraj University.

Research in applied fields of gonadal sex manipulation, cryo-preservation of sperms, potent hybrids and preservation of pure line stocks is leading to fast-growing and high-yielding fish varieties.

In the harvesting technology, designing and modification of various categories of trawl nets and fishing craft have improved fishing efficiency of artisannal and industrial sectors. In post-harvest sector, technologies for new and value-added products from unconventional fish species, chitin and chitosan from shell waste and prawn/fish flavours have been commercialised. Technologies are being perfected in the storage of frozen fish to increase shelf-life. A method for preparation of surgical sutures from fish gut has been devised.

The Central Institute of Fisheries, Nautical and Engineering Training (CIFNET) at Cochin, with a unit each in Madras and Visakhapatnam, aims at making available sufficient number of operators for deep-sea fishing vassels and technicians for shore establishments.

Integrated Fisheries Project, Cochin, envisages processing, popularising and marketing of unconventional varieties of fish.

The Central Institute of Coastal Engineering for fisheries, Bangalore, is engaged in techno-economic feasibility, study for location of fishery harbour sites and brackish water fish farms.

The ICAR has established eight different resource-specific fisheries institutes for developing technologies for exploitation of various resoureces from freshwater, brackish water, cold water, coastal and marine ecosystems, and for fish processing.

The Central Institute of Fishery Education is a deemed university which has promoted the development of trained manpower in fishery science, management and administration, harvest and post-harvest technology for both capture and culture of sectors.

Deforestation for the sake of other uses of land has been rampant, shrub jungles and fuelwood trees continue to vanish at an alarming rate. Environmental concerns have prompted planners to think of large scale planting of fast growing trees with suitable crops in between, i.e. combine agriculture and forestry. Tree-crop combinations have to be chosen carefully to minimise adverse effect on each other's potential. In the Gangetic plains, wheat or rapeseed in rabi and sesamum or pearlmillet in kharif can be grown in the interspaces of subabul trees. In the humid and sub-humid regions, maize is suitable for intercropping with Acacla and Stylosanthes for fodder for growing with Eucalyptus.

India is a vast land of diverse climates and varied soils. The increasing demand for food has necessitated agriculture to be practised in regions once considered "difficult". What makes these religions difficult is the lack of the right amount of one or more factors required for the usual crops—moisture or right temperature, for example.

The hot arid regions of India are spread over 31.7 million hectares. 61 per cent of them in West Rajasthan and about 20 per cent in Gujarat; the rest being in Karnataka and Haryana. The cold arid region is in Ladakh where aridity and low temperature limit agricultural season to about five months in a year.

In the cold arid regions some cereals, oil seeds and fodder crops which mature in a short period and can withstand severe cold are cultivated. The hardy pashmina goat can also be reared here.

In the hot arid regions, there is plenty of sunshine and high rate of evaporation. There is groundwater which should be scientifically tapped. Animal husbandry—cattle, sheep, goats—is possible in these areas. Fruit trees like ber and pomegranate and fuelwood trees like Acacia can be grown in these areas. Large scale planting of shelter belts will minimise soil erosion caused by wind.

One way to arrest the over exploitation of groundwater is to minimise the use of water, Low water duty crops like mustard can be taken up instead of high water duty crops like wheat. Use of sprinklers for arable crops and drip systems for high value (horticultural) crops is another step. There are positive indications that proper land treatment along with necessary mechanical structures, particularly on the ephemeral streams, can help in increasing groundwater recharge. This will lead to increased water availability besides diluting salinity.

Technologies are available for improving the productivity of the native pastures. These include reclaiming the degraded pastures with improved seed, contour trenching and planting of multipurpose trees (MPTs) such as Prosopis cineraria and fencing. Use of barbed wire has been suggested. With this system, the average productivity can increase four to five items from the base level of 500 kg/ha of forage production.

The Central Arid Zone Research Institute (CAZRI) at Jodhpur has laid out several demonstrations on strip cropping.

Rainwater can be harvested through ridging and furrowing of the surface or by providing one-third of the area as catchment for collecting run off for the remaining two-thirds.

Khadin are old rainwater harvesting systems, which have been in vogue since the 15th century in Jaisalmer district. Khadin is a practice which helps conserve water and soil. Use is made of rocky catchment areas to yield run-off water which is harvested in the valleys by withholding the rainwater with the help of a big bund. The water so collected is allowed to percolate and/or evaporate and by November winter crops like chickpea and/or wheat are grown on residual moisture. The system is stable and even in low rainfall years the khadins systain production. The productivity can be enhanced by choosing crops to match the stored moisture and using inputs like fertilisers to a moderate extent. Rela farming involves letting in water from the ephemeral streams on to the adjacent arable lands during the monsoon, impounding the same till November and then taking winter crops as in the case of khadins.

About seventy per cent of the 144 million hectares of arable land in India is rain-fed and depends on natural precipitation for crop production, and the practice of agriculture here is called dryland farming. Dryland agriculture is important for the economy; most of the cereals, pulses, oilseeds and raw cotton are grown on these lands. These lands receive rainfall between 500 and 1200 mm. It has been pointed out that even after the entire irrigation potential of India is developed by about 2000 AD about 50 per cent of arable land will remain rainfed. It is of utmost importance to increase the productivity of these lands.

Production in the rainfed areas can be substantially increased through appropriate soil and water management practices, including planned use of watersheds, water harvesting to provide for crop saving irrigation, soil moisture conservation and cropping systems with limited use of fertiliser.

The Indian Council of Agricultural Research (ICAR) initiated an All India Cordinated Project for Dryland Agriculture (AIRPA) in 1970. The Central Research Institute for Dryland Agriculture (CRIDA) was established at Hyderabad in 1985 with the main objective of developing practices leading to sustainable rainfed farming. Scientific projections help one to estimate the kind and amount of crops that can be harvested with the given rainfall. Several techniques of water management have been devised. Bunds — vegetative or otherwise — help to prevent runoff. Tillage renders certain soils more permeable and encourages high rate of infiltration. Premonsoon tillage is a foundation for rainwater conservation. Tillage also controls weeds and soil-borne pests. Contour or across slope cultivation boosts water intake. Drainage should, however, be kept in mind for black clay soils, for instance. Mulching helps to reduce water loss by evaporation. Seepage may be checked with cement-brick lining or even by plastering with dung and soil.

Dryland farming is risky, but the risk can be reduced by cultivating short duration high-yielding varieties of water — efficient crops. Water efficiency is improved by a moderate use of fertilisers containing nitrogen and phosphorus. Research has helped to design profitable and stable inter-cropping options to suit land and rainfall distribution. The CRIDA has recommended the following land use systems in rainfed areas :

1. Arable land : agro-forestry systems; intercropping with nitrogen-fixing trees species.
2. Cultivable waste and marginal lands : tree farming woodlots, range / pasture management; silvipastural management systems; timber and fibre systems.

Appropriate technology for agro-forestry and range / pazture management and allied practices to suit the prevailing agro-climatic conditions have been developed under the ICAR Coordinated Project on Dryland Agriculture and at the agriculture universities and research institutes in India. They could be adopted with great advantage to turn marginal lands into economical units.

The Government's aims for developmeent of dryland areas are (i) raising food production and smoothen out fluctuation in annual production; (ii) reducing regional disparities between irrigated and rainfed areas; (iii) restoring ecological balance by 'greening' rainfed areas appropriately; (iv) generating employment.

Rainwater moves down a slope, but it can be conserved effectively by dividing a geographical area into simple hydrological units. This is the basis of the concept of watershed, i.e. a piece of land that drains surface water through a single outlet, and the water-shed integrates the development of arable and non-arable areas. With the watershed as the unit of development it is possible to evolve a comprehensive approach on how best to use natural resources. The watershed approach was adopted in India with regard to rainfed agricultural in 1983 when a National Watershed Programme was piloted by the ICAR. The strategy was to combine the technologies on soil and water conservation with standard agronomic practices alongwith alternative land uses. The watershed programmes certainly led to some improvements; increase in crop productivity and cropping intensity, a shift in area in favour of more cash-yielding oil-seed crops from less remunerative coarse cereals, different farming systems based on contour vegetative barriers leading to moisture conservation.

Many watersheds — about 5000 hectares each in size — have been delineated across the country. The ICAR has provided expertise in tackling agricultural production and related soil and water conservation problems in 37 'model watersheds'.

In order to synchronise the activities of watershed development in the country, a full-fledged division of rainfed farming and watershed development and management has been constituted in the Ministry of Agriculture.

Hill farming inevitably brings to mind the age-old jhum, or slash and burn or podu cultivation practised by tribals. A patch of land is cleared on vegetation and plants burnt. The ash mixes with the soil. The crop is harvested, the land abandoned for another area. The land recovers fertility in five years or so. The system worked well in the days when production was small and shrub jungles plentiful. The land had enough time to regain its natural fertility. Today there is not much land available, the soils are given no time to recoup their nutrients losses in the natural way, no manure is applied, the held is low and soil erosion is rampant.

For controlling shifting cultivation and permanently settling Jhumia, a major scheme has been implemented in the north-eastern states. Activities undertaken were land development, irrigation, agriculture, horticulture, fisheries, social and fuelwood forestry.

Mountain agriculture though peculiar in many ways, is limited by a low land-to-man ratio — the cultivated land is only around 0.15 hectare as against 0.34 hectare in the country as a whole — and only about 18 per cent of the cultivated area has assured irrigation. Another major attribute of hill agriculture is that about 55 per cent of the cultivators are marginal farmers who possess less than a hectare. One more characteristic of hill farming system is that women make a major contribution in running the different operations. Mountain agriculture is further distinguished by complex and multiple farming systems such as agriculture, animal husbandry, horticulture, fisheries and agro and social forestry.

Although efforts to develop hill agriculture were initiated in the early Fifties, planned development programmes came to be evolved only in the late sixties. The concerted efforts made jointly by the agri-horticultural institutes and State development departments have certainly led to significant advances in agriculture and animal husbandry.

The main crops grown in the hills are wheat, maize, rice, barley, pulses, millets and vegetables including potato and ginger. Fruit crops grown comprise apple, cherry, pear, plum, peaches, apricot, walnuts, almond, mango and citrus species.

The introduction of Italian honey bee has also helped in boosting of the economy of the hills. Similarly mushroom cultivation has supplemented growers' income and the production per unit area has been sustained over the past decade. There has been significant progress in fish culture.

Based on the study of the slope and depth of the soil and availability of water, scientists have devised a suitable agricultural system; the upper reaches of hills are devoted to forestry, middle zone to fruit trees, perennial fodder grass and legumes, the third zone has terrace farms for a mix of crops. However, in practice, the system has not worked out very well.

For future, points emphasised are : (i) reduction of foodgrains crops and increase of horticulture and forestry, (ii) efficient use of run off harvesting, (iii) improved varieties, (iv) medicinal and aromatic plants should be encouraged, (v) fish culture, mushroom cultivation and bee keeping to be encouraged, (vi) evolution of technology for increasing the productivity of animals.

Farm machinery help to get the best results out of other inputs like seeds, fertilisers, water and plant protection chemicals. They also assist in increasing cropping intensity by reducing time for seed-bed preparation, harvesting and threshing.

Power operated equipment such as irrigation pumps, tractor and tractor-drawn machinery, and thresher have been very well adapted by Indian farmers. Along with tractors the sale of tillage implements such as cultivators, disc horrows and ploughs is also likely to increase. The equipment serve not only large farmers but also small farmers through custom hiring.

Improved animal drawn implements such as ploughs, horrows and cultivators were introduced among Indian farmers during the Fifties by simplifying the designs of foreign machinery. Improvement in indigenious tools and implements and development of new designs were initiated by the Indian Council of Agricultural Research through its institutes and agricultural universities. State departments of agriculture and laboratories of the Council of Scientific and Industrial Research (CSIR), Voluntary agencies and private entrepreneurs also contributed in this endeavour. Though the efforts were appreciable only a few designs could be got commercialised.

The assimilation of R&D requires an effective technological infrastructure of institutions and service to develop and test prototypes, to set up pilot plants for intensive evaluation and extensive demonstration, to set up standards besides a training infrastructure at all levels and credit support. No single agency can deliver the above and it requires perfect coordination among various promotional agencies.

With the establishment of the Central Institute of Agricultural Engineering at Bhopal under the aegis of the ICAR, the manufacturing and promotional activities were systematised. A network of regional R&D and demonstration unit was formed through all India coordinated schemes. These units developed technology for the small and marginal farmers. The State agroindustrial development corporations and private manufacturers were involved in the transfer of technology. A major thrust is being planned for the upgradation and adoption of proven technology through front line demonstration during the Eighth Plan period.

It is believed that the benefit of mechanisation could be reaped only by farmers having large acreage. The R&D institutions have developed simple low cost hand tools and animal drawn implements which even the small and marginal farmers could afford. What is required is creating an awareness among these groups of farmers about the utility of these machines through pilot introduction.

Agro-Industries Corporations and service centres work towards streamlining the system of input distribution and extending other agro-services to farmers. Farm machinery institutes at Budni (MP), Hissar (Haryana), Garladine (AP) and Biswanth Chairaili (Assam) impart on job training in selection, operation and maintenance of farm machinery.

The Department of Agricultural Research and Education (DARE) was set up in 1973 in the Ministry of Agriculture. It is responsible for coordinating research and educational activities in agriculture, animal husbandry and fisheries. It also brings about collaboration among various departments and institutions engaged in the same or allied fields on national and international levels. It provides government support, service and linkage to the ICAR.

The Indian Council of Agricultural Research (CAR) is a registered society, an autonomous body under DARE. It is the apex body responsible for promoting, conducting and coordinating research, education and primary extension education in agriculture, animal sciences, fisheries and allied sectors in India.

The country has been divided into 120 distinct agro-climatic zones and in each of them a multi-disciplinary regional research station has been established under ongoing National Agricultural Research Project.

The educational programmes are by and large conducted through 28 agricultural universities located in various States. Four of ICAR's institutes—Indian Agricultural Research Institute (IARI), New Delhi, Indian Veterinary Research Institute (IVRI), Izatnagar, National Dairy Research Institute (NDRI), Karnal and Central Institute of Fishery Education (CIFE), Bombay enjoy deemed university status and offer basic and postgraduate study programmes and award degrees.

Considering the growing importance of biotechnology in agriculture, the ICAR has established three national research centres in biotechnology—in agriculture, animal health and animal production—at three national institutes.

Agriculture is becoming knowledge intensive and is going to increasingly involve frontier technologies to accelerate research and development output. The priorities and thrust areas for future research are: (a) conservation and planned exploitation of germplasm resources (b) enhancing productivity through evolution of high yielding hybrids and varieties with tolerance to biotic and abiotic stresses. (c) development of integrated pest management practices for optimum plant protection, (d) breeder seed production, (e) research on export-oriented commodities, (f) diversification, with emphasis on agro-forestry, livestock and fisheries, (g) development and refinement of dry farming technology, (h) improving nutrient management system, (i) inventory of natural resources, (j) energy management, (k) post-harvest technology, (l) fostering excellence in research and education, (m) transfer of technology and improving communications system, (n) human resource development.

The ICAR performs the extension education functions and deals with first-line extension projects and activities. This involves demonstrating the latest technological advances made by scientists to farmers and extension workers, evolving suitable extension methods and serving as a feed-back mechanism for the research system. There are four first-line transfer of technology projects sponsored by ICAR.

For better transfer of technology, the ICAR has launched four programmes which are implemented by State departments of agriculture, State agricultural universities, ICAR institutes and voluntary organisations in some cases. The programmes are as follows:

(i) National Demonstration Projects (NDPs). The aim was to promote transfer of technology through demonstrations on multiple cropping, rainfed cultivation, special problem-soils and the entire farming system organised on individual farms. They helped the farm scientists get acquainted quickly with the problems of farmers in adopting new varieties and practices. The first demonstrations were held in 1964-65 on wheat, rice, jowar, bajra and maize. Later, the demonstrations were spread to cover the entire cropping system.

(ii) Operational Research Projects (ORPs). Launched during the Fifth Five Year Plan in 1975, the projects spread proven technology among farmers on a water-shed basis and study constraints to rapid dissemination of technical know-how including problems related to total resources development of the watershed area, and common agricultural problems.

(iii) Krishi Vigyan Kendras (KVKs). The KVKs or Farm Science Centres are grassroots level institutions. The KVK aims to influence productivity by imparting training through work experience and giving priority to the weaker social sections. Its mandate involves (i) collaboration with the specialists of the state agricultural universities and scientists of regional research stations for on-farm testing, refining and documenting technologies to produce sustainable and function specific land use systems, (iii) organising training to regularly update the extension personnel as well as training in agriculture and allied vocations to help in self-employment, and (iii) arrange frontline demonstrations to generate data relating to production, etc. with respect to various crops. (iv) Lab-to-Land Programme (LIP). This programme was started in 1979 to introduce low-cost relevant technologies that would aid in diversifying labour use and creating supplementary income in agriculture, fisheries, sericulture, animal husbandry, apiculture, rural crafts and other areas. Its objectives include ascertaining the resources and background of farmers to introduce low-cost, relevant technology; helping the farmers adopt feasible farm plans and improved technologies; and creating awareness among farmers of the agencies they can use for economic advantage.

In the technology transfer process, at present, 42 Central and National Research Institutes, four National Bureaux, Nine Project Directorates, 22 National Research Centres, 70 All India Co-ordinated Research Projects, 27 State Agricultural Universities, 126 National Agricultural Research Projects, Eight Trainers' Training Centres and 109 Krishi Vigyan Kendras are involved. Administrative sanction has been issued for 74 new KVKs.

The constraints may be technological, organisational and administrative, extension and social types, as well as some basic constraints. Specifically they relate to areas including marketing and pricing, soil and water resources conservation, evolving high-yielding and disease-resistant crop types, proper availability and use of chemicals such as weedicides, lack of transport and training facilities, and absence of proper study and in-depth evaluation.

To overcome the constraints and promote technology transfer, the following need to be emphasised:

- Appropriate technology development keeping in view farmers' needs and problems.
- Developing low-cost, profitable technologies that rely on locally available resources.
- Developing practices that involve 'optimum levels' of fertiliser, insecticides and other inputs so that small farmers can adopt them.
- Introduction of technology that is easily understandable where the farmers are concerned and which is also in keeping with previous experience and cultural belief and practices.
- Making coordination committees comprising representatives of development departments, input supply, credit and marketing agencies more active.
- Setting up district and state-level training institutions.
- Setting up linkage among research, extension and client systems.

In 1994-95 importance was given to the development of improved varieties and hybrids of different crops. The first variety of super rice in the world, (named 'Lunishree'), developed at the Central Rice Research Institute, has been commercially cultivated in coastal Orissa. The yield was increased, as the variety is salinity-resistant and can be grown on unirrigated low lands. Renu, MTU 933, Bipasa, and Amrut were new rice varieties for rainfed uplands, while for the rainfed lowlands there were Sagarsamba, Orugatlu and TPS-3. An integrated rice-fish-prawn-vegetables and horticultural farm system was developed with 5 to 15 per cent increased rice biomass production. High-yielding groundnut variety suitable for droughtprone area was identified for the first time. The IARI developed a new strain of genetically engineered mustard variety (Pusa Jaikishan), said to be the world's first genetically engineered mustard variety. It was the first attempt to somaclone (producing plants from test-cultured tissues) a food crop.

The IARI had identified a variety of man-made cereal. Triticale for rainfed conditions. A technology mission on maize has been initiated in 1995 to boost average productivity.

Biochemists in India have also created the first non-toxic strain of Kesari dal, a protein-rich pulse the sale of which was banned by the government because it contained neurotoxins that can cause paralysis. The new varieties are under field trials.

The IARI has also developed new varieties of carrot (Pusa Meghali) and cauliflower (Pusa Hybrid 2), High-yielding varieties of several tuber crops will encourage larger production for food and industrial purposes.

High-yielding pulses have also been developed and released for cultivation.

In the field of medicinal and aromatic plants, several improved varieties have been evolved. Agronomic practices including intercropping have been standardised.

Indian floriculture may develop as a commercial success with introduction of new ornamental crop varieties such as Arjun and Ratnagandha in rose, improved hybrids in orchids, and other plants.

Methods to improve the shelf-life of fruits-mango, guava and grapes have been standardised as have been methods for production of new products (e.g. Ber candy, jamun beverage).

In the field of animal science research, facilities for DNA finger printing and studies on major histocompatibility complexes have been developed. New strains of sheep — Avikalin, Avirastra and Avtmanas — have been developed for carpet wool production. High-yielding poultry layers have been developed, namely, ILM-90, ILR-90, Carigold, New feed resources have been identified.

Research in immunodiagnostics and vaccines has made advances. The virus responsible for equine coital exanthema has been detected for the first time in India. Cell culture vaccine against theileriosis has been evolved. Simple and reliable, DOT ELISA tests have been developed for diagnosis of various animal diseases.

In the field of fisheries, technology-packages for composite carp culture, air-breathing fish culture, integrated aquaculture with crop and animal agriculture alongwith use of organic waste has been of benefit to the rural sector. The maintenance and preservation of fish genetic material has been helped by cataloguing fish germplasm resources and successful long-term conservation of endangered fish species. Technologies for new value-added products from unconventional fish species have been standardised.

Women play an important role in agriculture. A National Research Centre for women in agriculture has been established at Bhubaneswar with the objective of generating specific technical information and refining it to make it available in suitable forms for women in all regions of India.

By definition sustainable agriculture is that form of farming which produces sufficient food to meet the needs of the present generation without eroding the ecological assets and the productivity of the life-supporting system of future generations. The term sustainability denotes the characteristic of a process or state that can be maintained indefinitely, and sustainable use of the ecosystem refers to making use of the system without impairing its capacity for renewal or regeneration.

The alternative techniques include crop rotation ; improved nutrient management — e.g. using legumes, animal manure, sewage sludge; tillage practices devised to prevent soil erosion — e.g. shallow ploughing deep subsoiling, ridge tillage; integrated pest management.

The shift from chemical to ecological agriculture should, however, be gradual. A sudden switch over could spell disaster and discourage farmers from taking to this course. At least seven years will be needed for the transition, and during the interim years the farmers could build up a sufficient organic base to fertilise the fields and improve the productivity of the soil, according to some preliminary studies of the ecological farms in South India.

Once a state of ecological balance has been attained and there are a good number of beneficial organisms to check the explosion of pests and pathogens, crop yields of a high order can be ensured. The cost of cultivation can be brought down substantially as many farm-grown inputs can be integrated efficiently in the farming system.

The adoption of ecological farming is not as simple as one may presume. It is a highly knowledge-intensive, labour-oriented and complex system integrating several organic recycling processes. A high degree of motivation, a firm commitment and innovative ability seem to make successful ecological farmers.

The integrated intensive farming system is a concept developed to meet the needs of ecology, equity and economics in a holistic manner. The system is being experimented in villages in Tamil Nadu under the joint efforts of the M.S. Swaminathan Research Foundation and the Association of Sarva Sewa Farms.

3

ATOMIC ENERGY

India has developed expertise in all stages of the nuclear fuel cycle, and created indigenous capabilities in designing, construction and operation of nuclear power plants, reprocessing of spent fuel, and manufacture of the necessary sophisticated equipment. Capabilities have been created right from exploration and mining of atomic minerals, preparation of high purity nuclear materials, such as uranium, thorium, ziroconium, etc., production of fuel elements for reactors, production of heavy water, health and safety instrumentation to nuclear waste management.

Though the country had a rich and varied tradition of fundamental research, organized studies in nuclear science began only in 1945, with the establishment of the Tata Institute of Fundamental Research. Thereafter, in 1954, a major part of the work on atomic energy was moved to Trombay, where a multi-disciplinary centre for research and development (the Atomic Energy Establishment, Trombay) was set up. In 1967, the centre was renamed as Bhabha Atomic Research Centre (BARC) in memory of its founder.

The prime objective of the atomic energy programme, as defined in the Atomic Energy Act of 1948, is the development, control and use of atomic energy solely for peaceful purposes, namely, the generation of electricity and the development of nuclear applications in research, agriculture, industry, medicine and other areas. To achieve this objective, efforts were initiated to build up a versatile infrastructure of research facilities, trained scientific and technical manpower, raw material processing centres and the know-how and capability to manufacture nuclear components and electronic equipment to support the atomic energy programme and make India truly self-reliant.

Thus, India started work on the peaceful utilization of nuclear energy at a time when it was essentially very much of a frontier science, into which a few developed countries alone had ventured. At a conference held in November 1954 in New Delhi on the Development of Atomic Energy for peaceful purposes in India, the late Prime Minister, Pandit Jawaharlal Nehru, in his inaugural address said, "so far as peaceful purposes are concerned, obviously we want atomic energy for the generation of power".

When the country's atomic energy programme was launched in the 1940s, a three-stage nuclear energy programme was envisaged to use the available uranium and vast thorium resources.

First Stage: natural uranium fuelled pressurised heavy water reactors (PHWRs) which would produce power, and plutonium would give uranium-233 from thorium.

Third Stage : once the size of the programme becomes large, sizable quantities of U-233 from thorium could be produced to start a series of thorium U-233 fuelled reactors.

The first phase has reached the commercial stage. The second phase has commenced with the successful operation of the fast breeder test reactor (FBTR) at Kalpakkam. Progress has also been made in the third stage with the successful development of a uranium-233 based fuel. Work has commenced on the design of an Advanced Heavy Water Reactor which will make the use of thorium in power generation.

India's long-term strategy is to return to thermal reactors for the third phase, since the thorium converted into U-233 would help to keep the cycle going, without sizable inputs of external fissile material. The energy potential of thorium in thermal reactors is way above that of natural uranium. India has nearly five times as much high-grade thorium as uranium, and the potential of thorium exceeds that of fast reactors. However, India has to depend on fast reactors for the first charge of U-233. The thorium-U-233 fuel cycle is of importance to India because of its nuclear fuel resource base. India has been making sustained progress towards the development of all aspects of thorium utilisation.

India has achieved technical capability in the design, construction and operation of PHWR of the first stage of the nuclear power programme. Self-reliance has been achieved in all segments of nuclear fuel cycle covering exploration, mining, fuel fabrication, designing, building and operation of nuclear power reactors, reprocessing and waste management.

In 1974, India exploded its experimental nuclear device. The peaceful nuclear explosion, as it is termed, had a plutonium device. Placed at a depth of 107 metres at the end of an L-shaped tunnel in the Thar desert near Pokhran, the device was exploded to yield an equivalent of 12 kilotons of TNT. The explosion produced a crater 10 metres deep and 47 metres in radius. No radioactivity is said to have been released into the surroundings by the explosion.

The executive agency for all activities pertaining to atomic energy in the country is the Department of Atomic Energy (DAE), which was set-up in 1954. Policies pertaining to the functioning of DAE are laid down by the Atomic Energy Commission (AEC) which was set up in 1948. The portfolio of DAE has all along been under the charge of the Prime Minister.

The activities of DAE are primarily in the area of nuclear power generation, research and development in atomic energy and in the industries and minerals sector. These activities are carried out by its constituent units, public sector units and by institutions which are given financial assistance by DAE.

The building of nuclear power stations was originally in the hands of a Power Projects Engineering Division. In 1984, this division was reconstituted as the Nuclear Power Board, converted later (1987) into the Nuclear Power Corporation of India Ltd. (NPCIL). The NPCIL is responsible for designing, constructing, commissioning and operating all nuclear power reactors in the country.

The Tarapur Atomic Power Station (TAPS) was commissioned in 1969, signalling the start of generation of electricity from nuclear energy. It has two 210 MWe-capacity boiling water reactors fuelled by enriched uranium. They were set up by an American company as turn-key projects. TAPS, that provides electricity to Gujarat and Maharashtra, has completed 20 years of commercial operation. China recently agreed to supply TAPS with low enriched uranium

(IEU). The first fuel consignment was received in Hyderabad on January 5, 1995. TAPS will operate on a fuel core consisting of LEU and mixed uranium-plutonium oxide (MOX). The Chinese sale nas been subjected to IAEA safeguards. Earlier, India's decision to subject the use of indigenous fuel for the plant to international safeguards was approved by the IAEA.

The original agreement on a safeguards was trilateral, involving the US, India and the IAEA. First, the US stopped supplies and then France, which provided enriched uranium under the framework of that agreement after the US stopped supplies. After that, the US did not allow India to acquire enriched uranium from any country and also unsuccessfully tried to claim ownership of the spent fuel. Though the plant had adequate fuel stocked up to keep it running for more than a couple of years, for long-term purposes India embraced the option of reprocessing the spent fuel which anyway is not easily given to storage.

The Rajasthan Atomic Power Station (RAPS) at Rawatbhatta has two prototype PHWR units of 220 MWe capacity each, using natural uranium as fuel. It supplies electricity to Rajasthan. The two units attained criticality in 1972 and 1981. RAPS also falls under international safeguards.

The Madras Atomic Power Station (MAPS) at Kalpakkam started commercial operation in 1984. The two PHWR reactors are indigenously designed and constructed — the first such in India. Electricity is provided to Madras.

The Narora Atomic Power Station (NAPS) in Uttar Pradesh has two PHWR units of 220 MWe each. These are standardised versions on which the design of all future 220 MWe PHWRs is to be based. NAPS provides electricity to Uttar Pradesh and Delhi.

At the Kakrapara Atomic Power Station (KAPS) a 220 MWe reactor has synchronised with the grid. With this the installed capacity for nuclear power generation in the country came to 1,720 MWe.

Future plans include four more reactors of 220 MWe being constructed — two each at Rawatbhatta and Kaiga (Karnataka). The design of the 220 MWe has been upgraded to 500 MWe, and six such reactors are to be set up at Tarapur and Rawatbhatta.

A significant feature of the Indian atomic energy programme is that it has all along been backed by a comprehensive R and D programme encompassing a wide range of multi-disciplinary activities relating to atomic energy. This includes fundamental research in basic sciences, disciplines like nuclear engineering, metallurgy, medicine, agriculture, isotopes, etc. Research is also being carried out in FBR technology and frontline areas like fusion, lasers and accelerators. All the organisations of DAE which are engaged in these activities can be considered to be one of the following categories, namely, R and D units, PSUs, industries and mineral (I and M) sector units, aided institutions or service sector units.

R & D work is mainly carried out at Bhabha Atomic Research Centre (BARC) at Trombay, Bombay, Indira Gandhi Centre for Atomic Research (IGCAR) at Kalpakkam, Tamil Nadu, Centre for Advanced Technology (CAT) at Indore, Madhya Pradesh, and Variable Energy Cyclotron Centre, Calcutta, West Bengal.

The Bhabha Atomic Research Centre (BARC) was set up in 1957 as the Atomic Energy Establishment Trombay and in 1967 was renamed as BARC in memory of its founder Dr. Homi Bhabha. It is the premeter national centre for research and development (R & D) work in nuclear energy and related disciplines.

APSARA was the country's first research reactor of one MW built indigenously in 1956. CIRUS, a 40 MW research reactor, was commissioned at Tarapur in 1960 for developmental work such as production of isotopes besides training and experimentation. PURNIMA I, a plutonium-fuelled reactor built at Trombay, was modified to PURNIMA II to use U-233 fuel in solution form. DHRUVA is an indigenous 100 MWe reactor commissioned in 1985 for research in advanced nuclear physics and for isotope production. At Kalpakkam is a neutron source reactor, KAMINI. There is a 5.5 MeV Van-de-Graff accelerator at Trombay.

Collaborating with the Tata Institute of Fundamental Research (TIFR), BARC has set up a 14 MV pelletron accelerator. It has also set up a beryllium plant and radio pharmaceuticals laboratory at Vashi, New Bombay. The Variable Energy Cyclotron Centre at Calcutta is a national facility for advanced work in nuclear chemistry, production of isotopes for various applications and radiation damage studies on reactor materials. The Seismic Activity Monitoring Station at Gauribidanur near Bangalore helps in detection and identification of underground nuclear explosions anywhere in the world. The Nuclear Research Laboratory at Srinagar and High-Altitude Research Laboratory at Gulmarg in Jammu and Kashmir conduct research in atmospheric physics and cosmic ray physics. The Radiation Medicine Centre at Bombay uses radioisotopes for diagnostic and therapeutic applications. Production facilities exist at Vashi, Bangalore and Hyderabad for radiopharmaceuticals. ISOMED is the radiation sterilisation plant at Trombay.

The activities of BARC are wide, encompassing such fields as physics, chemistry, engineering, metallurgy, fuel reprocessing, fuel fabrication, radioisotopes, waste management, electronics, instrumentation, lasers, biology, agriculture, food technology, radiation, medicine, etc.

Following the global interest in fabricating high temperature superconducting materials BARC also geared its R & D in this frontier discipline : it has succeeded in synthesising a single phase superconducting compound of bismuth lead-calcium-strontium and copper oxide with a T-120 K. BARC is also spear-heading research in cold fusion, yet another frontier area (work is also done at TIFR, IISC, IIT, Madras). From BARC's experiments, it emerged that the primary product of deuterium-deuterium reactions was tritium rather than neutrons. Studies have also been undertaken to ascertain the mechanism of cold fusion. With a view to develop intense sources of particles and radiation for plasma fusion research BARC continued its research on pulse power and particle beam technology.

India's tokamak, Aditya, was installed in 1989 at the Institute of Plasma Research in Gandhinagar, Ahmedabad. It is an indigenous effort from designing to commissioning. Aditya can generate plasma at 5 million degree celsius. The discoveries made by Aditya in plasma research and edge turbulence have had an impact on the world fusion research programme.

The Tata Institute of Fundamental Research (TIFR) Bombay is a national centre for advanced research in nuclear physics, mathematics, and high-energy physics and astrophysics.

The Indira Gandhi Centre for Atomic Research (IGCAR) is mainly concerned with FBTR technology : India's FBTR attained criticality in October 1985. The reactor core uses an indigenously developed sodium-cooled, mixed carbide (of plutonium and uranium) fuel. It is based on the design of the French Rapsodie reactor and is operating at around 1 MWe power level.

The FBTR's main purpose is to study the irradiation capacity of various materials including fuel and systems that go into a large liquid metal cooled fast breeder reactor.

The Nuclear Research Laboratory (NRL) of the Indian Agricultural Research Institute has used nuclear technique in delay the ripening of fruits and vegetables so as to increase their shelf life. Nuclear technique can also prevent damage to grams, lentils, mushroom, rice and cut flowers and thus enhance their trade and export possibilities. Nuclear magnetic resonance technique has also been used to identify the internally diseased alphonso mango in an apparently healthy lot. This is expected to go a long way in promoting export of alphonsos. On an average up to 25 per cent of alphonsos are diseased in each healthy lot. Nuclear science has helped to detect the actual use and wastage in fertiliser application to crops. Nuclear technique has also helped in groundwater investigation.

One of the main requirements for developing improved varieties of crops is the non-destructive mass screening process using low field nuclear magnetic resonance to identify and select desirable genetic material in plant breeding programmes. The NRL identified seven varieties of wheat for rainfed and four for irrigated areas. Nuclear technique has been used for determination of genetically superior oilseeds, without any damage to the seeds. Estimation of degree of saturation of oil in oilseeds and determination of rubber content in rubber later can be done within minutes by the nuclear magnetic resonance method.

Food preservation—be it drying, freezing, canning, sterilising or irradiation—aims at killing or inactivating insect pests, reducing the level of bacterial contamination or, in case of fruits and vegetables, delaying their ripening in order to increase their shelf life. Food irradiation, despite being a simple and safe technology, has not achieved the popularity it deserves. Reasons : (i) costly process (ii) technology, still not fully understood (iii) requires strict regulatory procedures (iv) public fear of anything to do with 'radiation'—notions that it will make food radioactive, induce chemical changes in it, and cause danger to workers involved.

Food irradiation programmes were taken up seriously only after 1981 when a blanket clearance was given by the joint expert committee of World Health Organisation, Food and Agricultural Organisation, and the International Atomic Energy Agency, that all foodstuffs irradiated up to 10 kGy (Gy stands for gray, which is a measure of the absorbed energy) could be used without any further toxicological or nutritional testing.

Process Food irradiation means exposing food to high-energy electromagnetic radiations such as X-rays and gamma rays of accelerated electron beams having energies up to 10 million electron volts (Mev). By penetration into the cells of microorganisms, the radiations cause structural and functional changes in biomolecules leading to cell death. The effect is either direct, that is, the molecules absorb radiation energy causing self-destruction, or indirect, that is, the radiation is absorbed by other molecules which react with the cell constituents. Investigations suggest that the DNA of the cell is the prime target of these radiations but the precise mechanism by which DNA damage occurs is still not clear.

Both X-ray /gamma ray and electron beam methods have their advantages and disadvantages. Electron beams are safer since they do not involve transport of radioactive material. They can be directed at the target and turned off for repair or maintenance work. However, the method is expensive and the electrons have a low penetrating power. So they can be used for irradiating a thin layer of food, for example, grain on a conveyor belt.

On the other hand, X-rays and gamma rays have a highpenetrating power, but the former are again expensive and not used in food irradiation. The choice today seems to be gamma rays, which require strict precautions since they are emitted in all directions from a radioactive source. The radio-active nuclides, cobalt-60 and caesium-137 have been selected internationally as gamma ray sources, taking into account their economics, availability and safety. Neither of the nuclides gives rise to any gaseous intermediates as they decay, which is important since the radioactive source is encapsulated in a stainless steel rod and welded.

All foods, of course, cannot be irradiated; milk, eggs and high-fatfoods, for example, cannot be irradiated. But irradiation can be very useful in other cases.

Irradiation followed by storage at 15°C enables long-term storage of potatoes and is a cheaper alternative to storing under refrigerated conditions. Onions also have to be stored under well-ventilated conditions to prevent their otting. While the actual dosage used depends on the function to be carried out, the energy of the gamma radiation is much below the level which can induce radioactivity, the main fear of the public. In case of food irradiation there is no direct contact with the radioactive source and the food does not get contaminated just as a bag passing through an X-ray machine does not become radioactive.

- A low dose (less than 1 kGy) is meant for sprout inhibition in bulbs and tubes, disinfectation of grains and grain products, delay in ripening of fruits.
- A medium dose (up to 10 kGy) is for preservation of fish, microbial decontamination of spices.
- A high dose (more than 10 kGy) is for sterilisation of meat products and medical products.

Use of chemicals leaves behind toxic residues. Present measures taken to prevent losses in stored grains due to insect infestation include fumigation with chemical pesticides. It has been shown that irradiation of wheat is effective against all stages of insect life. It must be emphasised, however, that irradiation alone will not be sufficient. In the absence of proper storage facilities and handling practices, reinfestation can occur. Evidence as to whether fungal infestations are affected by irradiation is confusing and conflicting.

As irradiation is a cold process and does not increase the temperature of the food, it allows the processing of frozen foods such as fish and shrimps. Due to the highly penetrating nature of radiations, the food can be irradiated in prepacked conditions, thus avoiding recontamination. It eliminates salmonella in frozen sea food. Considering that ethylene oxide is banned the world over because of its toxic residues. Irradiation at a maximum of 10 kGy — which is accepted as safe — would be ideal for our spices exported to the West.

Intensive research on irradiated food in the USA, the UK, France, Germany, India and China has brought to light many facts to prove that irradiated food is safe for human consumption. The

Chinese researchers had even fed irradiated food for a period of seven to 13 weeks to human volunteers, without any harmful effect to them.

Irradiation of food produces very little chemical changes in it. Many scientific studies using very sensitive analytical techniques have been done to detect the formation of radiolytic products, but no such products have yet been discovered. The loss of nutrients like vitamins during the irradiation process is comparable to the loss in other processing methods like heat treatment. The losses could be reduced if oxygen was eliminated and the temperature was reduced.

The National Monitoring Agency (NMA) constituted by the Government of India to oversee all aspects of food irradiation has accorded clearance for the irradiation of onions, frozen sea-foods and spices for export. The Union Government plans to permit trade in irradiated onion, potato, spices and frozen sea-foods for national and international markets, following amendments in the Atomic Energy Rules (Control of Irradiation of Food), 1991 and the Prevention of Food Adulteration Rules.

The IGCAR has developed several relatively cheap and highly sensitive electrochemical sensors to continuously monitor the purity of liquid sodium used as coolant in the fast breeder reactor. Liquid sodium is one of the critical technologies involved in fast breeder reactors. With a boiling point of 882°C liquid sodium remains highly subcooled even at a reactor temperature of 530°C. Consequently, the liquid sodium based primary cooling circuit can operate at near atmospheric pressure, while high pressure is required in conventional reactors to maintain the coolant water at temperatures of around 260°C. Being a metal, liquid sodium has much better thermal conductivity than water and can extract heat more efficiently. Its high reactivity has, however, raised concerns of safety, Sodium at above 200°C readily burns in air. More importantly, sodium reacts violently with water to produce hydrogen, with considerable potential for a dangerous explosion. Liquid sodium is a very benign substance when it is pure. Corrosion of stainless steel (used to make the reactor container and piping) is, for instance, far less than if water were to be used. It is therefore of prime importance that the purity of the liquid sodium be monitored continuously. IGCAR has accordingly developed cost effective electrochemical sensors for hydrogen, carbon and oxygen in liquid sodium. Plans are on for a 500 MWe prototype FBTR to be operational around the turn of the century.

The Centre for Advanced Technology (CAT) which was set up in 1984 at Indore, Madhya Pradesh, spearheads research in frontier technologies such as fusion, lasers and accelerators. It has constructed two synchrotron radiation sources and developed versatile lasers such as 70 W and 400 W carbon dioxide lasers for industrial, medical and research applications.

The Atomic Minerals Division (AMD) at Hyderabad is entrusted with exploration, prospecting and related development activities for identifying atomic minerals resources in the country. It acts in coordination with the Uranium Corporation of India Ltd. (UCIL) to establish the economic viability of uranium reserves at any given location. UCIL is responsible for mining, milling and processing of uranium ores. It operates mines at Jaduguda and Bhatin in Bihar.

Our uranium reserves are enough to support the first stage 10,000 MWe of nuclear power generation. Thorium reserves could provide support for 2,50,000 MWe. Besides Bihar, uranium reserves have been discovered by the AMD at Domtasat (Meghalaya) and Tummalapalle (Andhra Pradesh). Thorium is available in the monazite sands of Kerala's beaches.

UCIL supplies the end product (uranium concentrates) to the Nuclear Fuel Complex (NFC), Hyderabad for fabricating nuclear fuel for all nuclear plants in the country. Its special materials plant is a unique facility in the country for the production and supply of high purity materials required for electronics and other industries.

The Indian Rare Earths Ltd., has two mineral sand separation plants — at Manavalakuruchi (Tamil Nadu) and Chavara (Kerala). Its rare earths plant at Alwaye, Kerala, processes monazite in beach sands. It has also a project, the Orissa Sands Complex (OSCOM) at Chhatrapur, Orissa. The major products of the corporation are ilmenite, rutile, zircon, rare earths chloride and others.

Heavy water, the moderator and coolant for PHWRs, is produced indigenously. India has not only acquired self-sufficiency in heavy water production but acquired exports surplus also. The Heavy Water Board designs, makes and operates heavy water plants in the country. In 1962, the first heavy water plant was set up in

Nangal (Punjab). Plants using ammoniahydrogen exchange process came up at Vadodara which is in Gujarat, Talcher (Orissa), Tuticorin (Tamil Nadu), Thal (Maharashtra), and Hazira (Gujarat). At Kota (Rajasthan) and Manuguru (Andhra Pradesh) there are plants using hydrogen sulphide-water exchange process which has been developed indigenously.

The Electronics Corporation of India Ltd., Hyderabad, was set up to manufacture electronic systems, instruments and components developed at BARC required for carrying out research and development work. Subsequently, it has pioneered the production of a wide range of indigenously developed sophisticated electronic systems, instruments and components for use in industry and the nuclear power programme.

Recently there has been much apprehension about the dangers inherent in nuclear plants-fears of radiation hazard, waste disposal, disastrous accidents. While some of the Hazards are real, nuclear scientists point out that many of them are not based on scientific facts and unbiased observation. Furthermore, critics do not take into account the fact that nuclear plants are set up with great care for safety.

There is no doubt that radiation causes damage to living cells — but this depends on the intensity of radiation and the time of exposure. When an atom of a complex organic cell is exposed to radiation, ionization takes place and molecules disintegrate, adversely affecting the biological system, sometimes even destroying the cell. While high doses are fatal, low doses may have cumulative effect and cause cancers especially of the skin, and leukemia. It may affect lymphatic tissues, the nervous system and reproductive organs. However, the adverse effects take place on considerably high and constant doses of radiation. The release of radioactivity into air and water from reactors does take place but it is kept well within the limits prescribed by the Atomic Energy Regulatory Board. The earth is being constantly bombarded by cosmic ray nuclear particles (65% of natural radiation experienced by a human being is due to this). Background radiation from terrestrial and extra-terrestrial sources is much higher than radiation from nuclear plants. In the circumstances the radiation exposures from nuclear plants is of a negligible quantity. The fear of radiation arises because most people are unwilling to believe in any "safe level" for radiation exposure.

Recently a scare was created by a TV report that severe genetic diseases and congenital birth malformations have resulted as a result of radiation exposure to people in the villages even as far as 10 km from the power plant at Rawatbhatta. Nuclear scientists dispute this because so far there is no statistically significant scientific evidence of radiation-induced birth malformations or chromosomal abnormalities.

A controlled study of childhood leukemia around the Sellafield nuclear power plant in England claimed that there could be a link between the high incidence of childhood leukemia in Sellafield and high exposure does to the fathers employed in the plant. But this relates to doses much higher than normal environmental releases. And even this claim has been disputed on grounds of insufficient baseline data with regard to prevalence of clusters of such abnormalities in other parts of Britain.

Another aspect of nuclear hazard is waste management. The general technique in dealing with radioactive wastes is to concentrate and contain as much radioactivity as possible and discharge to the environment only effluent of as low a concentration level as possible. At inland sites like Narora and Rawatbhatta, low level liquid wastes are discharged into the environment at a minimum level. At coastal sites such as Tarapur and Madras significant dilution in the sea is possible. For solid wastes, different types of containments are used and located at sites selected on the basis of geological and geohydrological evaluation.

The highly radioactive wastes from fuel reprocessing are carefully managed by incorporating them in inert solid matrices, placed in canisters which are kept under cooling till the radioactivity comes to desired level. Finally the canisters are stored in suitable geological media.

A waste immobilization plant for incorporating the high level radioactive wastes generated from the fuel reprocessing plant is set up along with the solid storage surveillance facility at Tarapur. Immobilization involves verification of radioactive waste which is coded at underground disposal.

India happens to be among those few countries in the world which have developed the technology to convert high-level radioactive waste into glass.

The sites for nuclear plants are chosen with safety parameters in mind. The plants are designed for safe operation and take into account potentially serious accidents through a series of protective measures. Recognising the possibilities of human error, equipment malfunctions and extreme natural phenomena, the plants are designed on the concept of 'defence-in-depth'. There are multiple physical barriers which would have to be successively breached for any significant amount of radioactivity to escape from the reactor. If these precautions fail, further measures are prescribed to cope with the emergency.

As for radiation, the amount is carefully limited to much below what we experience from natural sources, or X-ray. Even the plant personnel are given the maximum protection possible. There has been a significant improvement in the design, construction and operation of power reactors in the country, as a result of which radiation levels have progressively been reduced.

The external radiation doses received by occupational workers from all over the country are monitored on a monthly basis. Film monitoring service is provided to people working in medical, industrial and research institutions. Thermo-luminescent distmeter monitoring service and fast neutron monitoring service are provided to the people working in reactors, fuel reprocessing plants and accelerators. Indeed, the Atomic Energy Regulatory Board has reduced the effective dose limit of radiation workers are subject to in nuclear, medical, industrial and research installations from 35 millsievert (MSV) to 30 MSV for the year 1993. This marks a progressive step in implementing the latest recommendations of the International Commission on Radiological Protection (ICRP). It may be recalled that the ICRP in 1990 recommended for radiation workers an effective dose limit of 20 MSC per year average over five years with the further provision that the effective dose should not exceed 50 MSV in any one year. In 1991, AERB restricted the dose limit of radiation workers to 40 MSV. In 1992, the value was further reduced to 35 MSV.

The Atomic Energy Regulatory Board (AERB), an autonomous body reporting to the AEC, carries out regulatory and safety functions as envisaged under the Atomic Energy Act, 1962, covering all DAE units and other radiation installations operated by medical, industrial and research users in the country. It is empowered to frame rules and regulations related to safety and regulatory requirements. It is also empowered to specify safety standards, codes and guides as envisaged under the Atomic Energy Act, relating to site-selection, design, construction, commissioning, operation and de-commissioning of all nuclear installations. In addition, it has authority and responsibility to ensure compliance with standards and codes in both DAE and non-DAE organisations. The board is assisted by a number of committees with eminent scientists and engineers from industry, universities, etc. as members. Commissioning of any nuclear installation, be it a power reactor or a small research reactor, has to be approved by the AERB.

In 1987 a Crisis Management Group was formed, and it is responsible for ensuring that each of DAE's nuclear installations is adequately prepared and equipped to deal with any possible emergency involving radiation or conventional hazards at the installations or during transportation. An on-going programme updates emergency preparedness plans at all nuclear installations.

On the whole, India's safety standards are in keeping with those laid down by the International Atomic Energy Agency (IAEA) and the International Commission on Radiological Protection (ICRP). As a part of a global network to monitor environmental radiation, India is also establishing a national network of monitoring stations to detect unusual radiation releases. Stations are already operational at Tarapur, Mumbai, Kalpakkam, Calcutta and Indore. "SODAR", the sophisticated weather monitoring system operates from Kalpakkam and the Kaiga collecting meteorològical data of the site. An environmental dose logging system has been developed for use in the vicinity of nuclear power plants.

The AERB in its March, 1996 report rules out a Chernobyl type of catastrophe in India's nuclear power plants, but calls for improvements in the emergency core-cooling systems of the reactors. Proper inspection and evaluation of the zircalloy pressure tubes of the PHWRS are also recommended. Evaluation of structural response to earthquake forces in some facilities and preventive measures related to fire and flood have also been called for.

The IAEA classifies events on the International Nuclear Event Scale—a scale of 0-7 depending on their severity. Events which may be termed 'accidents'—level 4 and above on the scale—have all so far happened in the West (Chernobyl was 7 on the scale; the Narora fire was put at level 3). What is more, weapons complexes have a far greater degree of safety related problems.

The radiation fields that develop in the out-of-core surfaces of pressurised heavy water reactors (PHWRS) pose a problem to personnel involved in the maintenance of the nuclear plants. The way out is to 'decontaminate' the plant every few years. Abroad, it is done every five years or nearabouts. The first unit of the Madras Atomic Power Station (MAPS-I) at Kalpakkam, was recently decontaminated using chemicals specially developed and standardised by the Bhabha Atomic Research Centre.

The technology developed by BARC to extend the life-span of coolant channels in nuclear reactors makes India the second country in the world to have acquired this capability. A special system called the coolant channel replacement machine (CCRM) has been devised. A new alloy with an expected life of 30 years has been developed for this purpose. MAPS-II became the first operating Indian reactor to utilise the indigenously developed technology.

Nuclear science technologies developed by BARC have been transferred as 'spin-offs' to the industries for commercial exploitation. The health of Indian Oil Corporation's pipelines can be checked by a pipeline inspection gauge. The Thyristor Chopper Control systems has helped the Railways to reduce power losses on suburban trains. Recently studies have been undertaken on silt movement in new Mangalore Port and Hooghly River estuary, flow measurements in River Teesta in Sikkim, and water seepage identification in Chaskaman Dam in Pune.

Government has approved irradiation of foodstuffs (spices, potato and onion) for international consumption. Improved crop varieties include groundnut and black gram; banana plant has been produced by tissue culture.

Radio-immunoassay kits have been developed and distributed to many institutions for early diagnosis and treatment of diseases like cancer, Gamma sterilisation plants have been established in many parts of the country. The medical laser units manufactured by the Centre for Advanced Technology are being used for surgical applications.

A supercomputing system, Anupam, based on parallel processing, an intelligent Braille interpreter system, antenna systems, advanced remote handling and robotic devices are other spin-offs of nuclear technology.

4

BIOTECHNOLOGY

With its strong background in the traditional areas of plant, animal and medical sciences, India is now poised to become the first developing country in this part of the world to make use of biotechnology to solve many problems being faced in medicine, agriculture and industry after breaking big powers' monopoly in nuclear energy, space and deep sea mining.

In recent years, biotechnology, which involves use of novel microbes altered or manipulated by man through genetic engineering techniques, has emerged as an important discipline. It can be broadly defined as the development forms and systems for obtaining maximum benefits to man and other forms of life while maintaining an optimum ecological balance by the use of conventional and non-conventional techniques of interaction between the living and the non-living as well as between the living and the living.

Over the years, man has exploited biological processes in various technologies such as fermentation of alcoholic beverages, food processing and preservation, retting of fibres, curing of tobacco, production of vaccines, pharmaceuticals and other chemicals any recycling of industrial and animal wastes. The recent epochmaking discoveries which have led to the development of techniques of genetic engineering, cell-cell fusion amongst plants and animals and microorganisms and manipulation of enzyme and metabolic pathways have opened up new possibilities for the cost-effective production of fuel from renewable resources, cloning and better yielding varieties of plants, low cost medical testing kits, better methods for early diagnosis of selective methods of pest control and purification of waste water. Several basic disciplines of science like biochemistry, physics, mathematics, chemistry and engineering have an inter-face with biotechnology.

In 1982 Government of India set up the National Biotechnology Board which was replaced by a separate Department of Biotechnology (DBT) in the Ministry of Science and Technology in 1986. This was in recognition of the need for a focal point in the administration and structure of the Government for planning, promotion and coordination of biotechnological programme.

Realising the need of properly-trained manpower for supporting programmes in multi-disciplinary and fast moving areas of biotechnology, the Department of Biotechnology has evolved a very strong integrated manpower development programme, including a model system of post-graduate and post-doctoral teaching in biotechnology, in selected universities/institutions all over the country in collaboration with the University Grants Commission, Ministry of Human Resource Development, ICAR, ICMR and Department of Ocean Development; short-term training courses to impart practical training in various newly-emerging research techniques to meet the immediate requirements of expanding R and D activities in biotechnology within the country; awarding biotechnology associateships every year for conducting advanced research or undergoing specialised training in identified priority areas of biotechnology within the country and abroad. The Department of Biotechnology, has created a network of infrastructural facilities for building up a strong R and D base and manufacturing capabilities. The facilities which have been set up include germplasm banks for plants, animals, algae and microbes, animal house, oligonucleotide synthesis, production and distribution of enzymes, reagents and radio labelled compounds, bio-process optimisation and pilot plants, etc.

Two autonomous institutions have been established, namely, National Institute of Immunology (NII) at New Delhi and National Facilities for Animal Tissue and Cell Culture (NFATCC) at Pune. NII is concerned with the control of fertility and the diagnosis and control of communicable diseases. Its main areas of research are in birth control vaccines, immuno-diagnostic kits, predetermination of sex of embryos, aquaculture bio-technology. The NFTACC is concerned with maintaining, identifying, storing, propagating and supplying human and animal cell lives.

The DBT has constituted an expert committee on conservation of biodiversity under its plan to conserve endangered and threatened plant species which are economically and medicinally important. For in vitro conservation of germplasm the ICAR has established a National Facility for Plant Tissue Culture Repository (NFPTCR) at the National Bureau of Plant Genetic Resources, New Delhi. The NFPTCR uses tissue culture and cryopreservation of plants. Gene pools can be preserved for use in breeding superior plants. The DBT has set up national gene banks at the Central Institute of Medicinal and Aromatic Plants (CIMAP), Lucknow, the National Bureau of Plant Genetic Resources (NBPGR), New Delhi, and the Tropical Botanical Garden and Research Institute at Trivandrum. These banks will conserve medicinal and aromatic plants extensively used in traditional systems of medicine.

Biotechnology products being highly skill-based and not high investment-oriented, linkages between R and D institutions and industry need to be established with the clear aim of commercialisation. It was therefore decided to set up Biotech Consortium (India) Limited. The scope of the Consortium would be transfer of technology, product development, feedback from market, minimising investment of risk capital as also to supplement the efforts of existing institutions.

A national network of Distributed Information Centres (DICs) and Distributed Information Sub-Centres (DISCs) in specialised areas of biotechnology has been established under DBTs Biotechnology Information System (DTIS) programme. The present scope of the programme covers several aspects, e.g., data capture technologies, data banks, mathematical models, artificial intelligence/interactive software and access to all bibliographic information on biotechnology. Its purpose is to provide a national bio-information network designed to bridge the inter-disciplinary gaps on information and to establish links among scientists in organisations involved in R and D manufacturing activities in biotechnology; to build up information resources, prepare databases on biotechnology and to develop relevant information handling tools and techniques; to establish linkages with national and international databases and information centres on commercial terms or on exchange basis to evolve and implement programmes on education of users and training of information scientists responsible for handling of biotechnology information.

Bio-technology has a remarkable variety of applications for the benefit human beings and their environment.

The Department of Biotechnology is supporting several R&D projects in the area of crop biotechnology, such as regeneration of plants from protoplasts of indical rice and wide hybridisation.

Plant Molecular Biology work is being conducted at the Tamil Nadu Agricultural University, Coimbatore, Madurai Kamaraj University, Madurai, Jawaharlal Nehru University, Delhi, National Botanical Research Institute, Lucknow, Osmania University, Hyderabad, and Bose Institute, Calcutta. These centres shall have the responsibility for both research and training. They will see to (a) infrastructure strengthening for conducting basic research in plant molecular biology and (b) initiating three to four highly focussed crop-based research programmes on molecular aspects related to plant tissue culture, RELP mapping, chloroplast and mitochondrial DNA research, among other things. Isolation and characterisation of genes encoding seed storage proteins from grain amaranthus has been achieved. A nutritionally important gene encoding lysine-rich protein has been cloned.

Recently, researchers at Guwahati University succeeded in making regenerated plantlets of a rare species of orchid to flower in a test tube in just eight months. It was part of the project "in-vitro propagation of endangered species of north-eastern India." A breakthrough was also achieved in producing planting material for pineapple by culturing the auxiliary bud of the fruit in a test tube.

Agricultural productivity and diversity can benefit from biotechnology. The impact of the following developments merit attention.

Biofertilisers. Instead of increasing the use of chemical fertilisers, it is being increasingly realised that the same function could be achieved by improving and transferring nitrogen fixation ability in plants. In this context, micro-organisms such as rhizobium, azobacter, clostridium and blue algae as nitrogen fixers, bacillus subtiles and micrortigal fungi as phosphate solubizers; azolla as organic matter contributor and nitrogen fixer and throbacilles as sulphate producer hold great promise. It has been found that inoculation with rhizobial cultures in different pulse crops increases yield significantly.

In India, the National Facility for Blue Green Algal Collection at IARI has come up with an inexpensive algal biofertiliser technology for rice. Higher grain yields have been demonstrated in field trials in various regions. Large-scale algal production has been undertaken at state seed farms, experimental stations and by individual farmers.

The Department of Biotechnology has launched a Technology Development and Demonstration Project on Biofertilisers, which envisages large-scale demonstration at farmers' field and also a continuing feed-back mechanism to improve the product. The main objective of the Blue Green Algae Project is to develop a strong R and D base for large-scale application of algal biofertilisers through standardisation and process optimisation for algal biomass production, genetic manipulation of algal strains for ecologically and agronomically useful characteristics, development of protocols for quality assurance and retrieval and standardisation of inocula. The project has been implemented at six centres in the country.

The project on rhizobium has been taken up in two phases. The main activity in the short-term will be process optimisation, production formulation and testing. The rhizobial strains will be supplied by the National Facility for Rhizobium Germplasm Collections at IARI, New Delhi (supported by DBT). The long-term objective would be : selection of suitable strains of rhizobium for different crops, strain improvement, identification and use of beneficial micro-organisms, improving shelf life in different carriers, quality control, methods of inoculation and development of a mega strain for different pulses and oil seed legumes. The project is being implemented by IARI, New Delhi and Annamalai University, Madras. There are different collaborating agencies involved.

Bio-insecticides and Bio-pesticides. Intensive and recurrent use of some of the chemical pesticides leads to toxicity in the environment and pesticide resistance in target pests. Biotechnology offers several means of overcoming the deleterious side-effects by allowing farmers to adopt alternative strategies toward pest control. These include the use of biological control agents, attractants and growth affecting agents. Besides being environmental friendly bio-pesticides are non-carcinogenic and have a high degree of insect specificity. A common bio-insecticide is a toxic chemical (protein) produced by a soil-borne bacterium, Bacillus thuringtensis (BT).

The use of BT was banned in India till 1991 as its spores were considered a threat to silkworm. However, now the NRCPB has developed non-sporulating strains of BY which are insect specific in toxicity.

The Department of Biotechnology has established a Control Network Programme with several R and D sub-projects throughout the country for the control of serious insects, pests, diseases of sugarcane, cotton, chickpea, pigeonpea, oilseeds and vegetables. The main objective of this network project is to lay greater emphasis on developing formulation and pilot scale agents in the farmer's field, and also for integrated pest management of key pests and diseases. Some biocontrol agents which were mass produced and whose efficacy for controlling the pests and diseases of various crops was demonstrated in farmer's field at multi-locations are: baculoviruses / microbial category: NPV or heliothis armigers; parasites and predators: (i) Chryosopa (for control of aphids, white files and bollworm and (ii) Trichoramma (for control of lepidopteran pests). Two biological control pilot plant units at Madurai and Coimbatore for mass production of biocontrol agents. The Biotechnology Consortium of India (BCIL) will assist in commercialising biopesticide technology.

Tissue Culture. Out of the various biotechnological techniques for agriculture, tissue culture is the most promising one. The technique involves taking out a piece of growing tissue of a plant, disinfecting it and culturing it in a suitable medium so that the mass of the cells begin to reorganise into the whole plant, a tiny facsimile of the original. This implies that from a single plant possessing all the desired characteristics, virtually unlimited ones can be propagated. Such micropropagation has been carried out in several vegetable, fruit and flower varieties. The advantage of the tissue culture technique is the propagation of true progenies ensuring uniform growth and productivity behaviour for each species in given agroclimatic environment through successive generations. The somaclonal variations arising through tissue culture cycle can be effective in bringing about genetic improvement in plants. Bio 902 is a somaclone of mustard developed by the NRCPB of the IARI.

Hybrid Seeds. Hybrids are the product of the union of two different geno-types and capable of performing their functions more efficiently than their parents under different agronomic conditions. This is usually done by biotechnological methods.

Artificial Seeds. Development of artificial or synthetic seeds consisting of tissue culture embryos encapsulated in protective coatings is an emerging area for the production of unlimited quantities of quality seeds.

Photosynthesis Improvers. Photosynthesis improvers are a series of new naturally occurring chemicals which have the potential for advancing plant productivity. Scientists have established that even I part per million (ppm) dosage can increase photosynthesis in plants by 100% or more and increase the yields of crops such as cereals, pulses and vegetables.

Stress Resistant Crops and Plants. New biotechnological techniques have enormously increased the scope for evolving improved varieties of plants of all kinds. It is now feasible to develop and fast propagate plants that are resistant to different constraints such as drought, salinity and various pests and diseases.

Animals meet the milk, meat, fibre and drought power requirement of the country. The application of biotechnology in this area in increasing production efficiency through manipulation and control of physiological systems and improving the health and well-being of animals assumes a very great significance.

With a long coastline and 4.5 million hectares of water area as ponds, tanks and lakes, aquaculture potential of India is yet to be harnessed fully. The potential areas of application of biotechnology in Indian fresh water aquaculture are recycling of organic matter, biological nitrogen fixation and genetic transfers.

Demonstration of technology feasible and commercially viable semi-intensive prawn farming with the application of biotechnology has been launched with the sponsorship of MPEDA (Cochin) and with the technical know-how of HLL and abroad.

A research and development project on the demonstration of intensive carp culture for achieving the production levels of 25 tonnes per hectare per annum has been implemented at CIFA, Bhubaneshwar. Under this project, certain biotechnological innovations such as chromosome manipulation, hybrid production, high energy high-protein feed formulation, fish disease diagnosis and control and overall work on pond management are being carried out to develop technological packages to achieve much higher production.

Under the Department of Biotechnology Project for Embryo Transfer, the infrastructure has been created throughout the country and appropriate skill pool has been generated for embryo transfer laboratory has been established at Bidaj, Gujarat, which provided the central facilities for the project for training, research and coordination. Four Regional Centres have been set up at Nasik, Nakerikallu, Rai Bareilly and Hassarghat. The collaborating agencies NII, IVRI, NDRI and CFSP and T.I. provide the research support in areas of cryopreservation, in ultro fertilisation, micro manipulation cloming and gene transfer for field application.

The tissue culture technique for mass multiplication offers means of not only rapid and mass multiplication of existing stocks of germplasm for woody and biomass energy production but also for rapdi afforestation of degraded forests and for regeneration of green cover.

Biomass is the utilisable source of carbon for biotechnological processes. Highest proportions of biomass is produced in forests followed by grasslands and agriculture.

The technique also called gene splicing and recombinant DNA technology, whereby, a section of DNA, comprising one or more genes, is removed from a cell and recombined with the DNA of another cell. The receiving organism is said to be transgenic.

Genes are the hereditary material that determine what a living thing will be — how it will develop and function — and what its descendants will be. Through recombinant DNA technology, scientists alter both the organism they work with and the generations that follow it. For example, genetically engineered bacteria now produce human-sequence insulin for use by diabetics.

Much recombinant DNA work is done with bacteria. In most organisms the DNA is contained in bodies called chromosomes. In bacteria, a small amount of DNA is also held in rings called plasmids, and it is these that are used in gene splicing. One often used bacterium is Escherichia colt (E. coli) an organism normally present in vast amounts in the human intestine. Bacterial produce substances called restriction enzymes. Each enzyme is specific — a king of molecular axe that cleaves a DNA molecule at a particular place but no other. (These enzymes normally operate against viruses that prey on bacteria). Another molecule, DNA ligase (Latin ligare, to bind), acts like a glue to hold together freshly joined sections of DNA.

By 1988 hundreds of human genes had been engineered. Some engineered substances human insulin. INTERFERON, for use against viruses and some types of cancer, and a vaccine for one form of hepatitis. Human growth hormone (somatrem) was available. Scientists were producing and testing genetically engineered vaccines against malaria and AIDS, as well as engineered products to treat heart attacks hypertension, and haemophilia.

Genetically engineered tissue plasmtnogen activator (tPA) has been used with great success to dissolve blood clots in the arteries of the heart, the cause of most heart attacks. Although tPA is produced normally by the body, the amounts needed for treating heart attacks are available only through genetic engineering.

Also called genetic fingerprinting, the technique is used for identifying the components of DNA (the material of the genes) that is unique to a particular individual. Just as one person's fingerprints are different from everyone else's and can be used for identification, so a small section of the DNA of an organism (which is present in every cell of the body) uniquely distingushes that particular organism from all others. These varying bits of genetic material take the form of sequences of DNA, called mini-satellites, which are repeated several times. The number of repetitions of a mini-satellite per region of a gene can vary enormously between unrelated individuals. Chemical analysis of an organism's DNA from a sample of blood, tissue, semen, etc. using the techniques of chromatography and electrophoresis, produces a two-dimensional pattern of spots. This corresponds to the genetic profile of that organism, complete with the repeating sequence which can be picked out.

Although comparatively new DNA fingerprinting is becoming an established forensic technique and has been used successfully as evidence in court cases of rape. It is also being used to investigate family relationships in animal populations, and to measure the extent of inbreeding by looking at the degree of variability in DNA profiles. Comparing the DNA patterns of parents and off springs may in future be useful in preventing trade in endangered species or in proving paternity suits.

Indian scientists have developed a universal DNA probe that can be used in a variety of instances.

Fourteen nationally important forest tree species have been identified for undertaking work on production of biomass fuel, fodder, timber and commercial / industrial woods using tissue culture technology. Two pilot plant units have also been set up at the National Chemical Laboratory, Pune and Tata Energy Research Institute, New Delhi to carry out large-scale micropropagation / bulk multiplication of elite forest tree species using tissue culture techniques with a production potential of a few million plantlets per year.

Biotechnology has already evolved methods for using aerobic as well as anoerobic microorganisms for purification of industrial wastes and sewage. Microbiological strains could be isolated in order to control various forms of toxic pollutions. Bacterial leaching in mining and use of micro-organisms or compounds made by microbial processes to coax entrapped oil from abandoned oil wells and oil shales are other areas of great commercial potential. Indian Organisations like BCIL, SPIC etc. have developed bioprocesses using microorganisms to deal with effluents from various industries.

Bio-technology offers economical process steps in producing various industrial chemicals. The Institute of Microbial Technology (IMTECH), Chandigarh and the Vittal Mallya Scientific Research Foundation, Bangalore, have achieved a significant breakthrough in the field of alcohol production from molasses using an improved yeast strain obtained by genetic manipualation. India has also developed the first biotechnology-based process in the world for invert sugar syrup production. The process developed by the Food Technology and Enzyme Engineering section of BARC has been licensed by the National Research Development Corporation (NRDC).

Use of biotechnology in improving human health has far reaching consequences. Areas in which impact could be felt are given below:

Vaccines Tissue culture based vaccines now could be produced in mass scale. These vaccines are highly effective and thermostable. Due to new techniques used in production, it will be able to retain its potency in near normal conditions. The DBT has promoted the establishment of two R&D cum manufacturing units for cell culture based human viral vaccines. Attenuated polio-vaccines, plasma derived Hepatitis B vaccine and other immunologicals are to be produced by the Bharat Immunologicals and Biologicals Corporation Ltd. at Bulandshahr, Uttar Pradesh. The development and manufacture of measles, vero rabies and vero cell-based oral polio vaccines are entrusted to the Indian Vaccines Corporation Ltd., Gurgaon. Effective and reversible contraceptive vaccines are being researched into and vaccines for men and women are undergoing clinical trials.

Diagnostics Biotechnological techniques are increasingly being used for effective diagnosis of diseases. Though at present the diagnostic kits are mostly imported, the Government is promoting research in this field at several institutions. Techniques for detection of pregnancy, fliariasis, typhoid, malaria and blood grouping sera have been developed.

Medicines Recombinant DNA technique is expected to revolutionalise the production of protein pharmaceuticals — insulin, growth hormones, interferons, blood-clotting factors.

Gene therapu : In genetic disorders the only cure appears to be the replacement of the defective gene with a normal gene. One of the diseases considered for gene therapy is severe combined immune deficiency.

5

HEALTH

India's major medical research centres are making use of advanced technologies to find solutions to the country's health problems. Scientists are using conventional as well as modern tools of medical sciences like genetic engineering and biotechnology to cope with the present and oncoming health problems like AIDS, tropical diseases and fertility control.

At the forefront of these developments are efforts to develop vaccine against tropical diseases like malaria, leprosy, as also anti-fertility vaccines and high-tech babies. India's first biotechnology product—a pregnancy kit—has already entered the U.S. market. And India's first anti-fertility vaccine is undergoing trials in India and abroad.

Research in medical sciences was started by a few British medical men on their own or as government officials as they found that India, with a wide prevalence of communicable disease, was a fertile field for medical research. The important fields studied were the role of anopheles mosquitoes in the transmission of malaria by Ronald Ross; the development of vaccine for the prevention of plague by Haffkine; studies on trypanosomes filaria by Lewis; investigations on spirillum of relapsing fever by Van Dyke Carter; cholera by Macnamara and Cunningham; and snakes and snake venoms by Fayrer. A string of Pasteur Institutes began to be established from 1900 onwards. Donovan working in Madras Medical College demonstrated the presence of a parasite, christened as Leishmania Donovani, in the spleens of kala-azar patients. Among

Indian scientist, Brahamachari discovered Ureastibamine in 1920 as a specific treatment for kala-azar. Rogers demonstrated the effectiveness of emetine in the treatment of amoebiasis.

In 1911, the Government of India established the Indian Research Fund Association to promote scientific medical research in India. This Association, for over 40 years, sponsored schemes of research on malnutrition, malaria, rabies, plague, cholera, leprosy and kala-azar. Some of the centres where such work had started developed into institutes.

At the time of independence, health services in India were rudimentary. Epidemics of smallpox, cholera, malaria and plague were frequent and fatal. The crude death rate was about 17/1000 population and infant mortality 134/1000 live births. Expectation of life was 31 years. Over the years, the situation has improved through the implementation of various plans and policies. The Bhore Committee Report submitted in 1946 itself pointed out the necessity of an integrated primary health care, social and preventive medical care, people's participation, and strong intersectoral cooperation.

The Indian Constitution lays down as goals to be achieved—"elimination of poverty, ignorance and ill-health". It directs the State to raise the level of nutrition and improve public health, secure the health of workers, men and women and ensure that children are given opportunities and facilities to develop in a healthy manner.

India, a signatory to the Alma Ata declaration of 1978, is committed to the goal of "Health for All" by 2000 AD. It is with this commitment in the background that the National Health Policy was adopted by Parliament in 1983, which is a blueprint for action by the Government and voluntary agencies. The policy stresses upon the preventive aspects of health care besides the need of establishing comprehensive primary health care services to reach people in the remotest areas of the country. It also views health and human development as vital components of overall integrated socio-economic progress. It calls for a decentralised system of health care and people's participation.

Under the Constitution, public health and sanitation and hospitals and dispensaries fall in the State List. Population control and family planning, medical education, adulteration of food stuffs and other goods, drugs and poisons, medical profession, vital statistics including registration of births and deaths, lunacy and mental deficiency are in the Concurrent List.

The Union Ministry of Health and Family Welfare plays a vital role in the national efforts to help citizens lead a healthy and happy life. The Ministry is responsible for implementing programmes of national importance like family welfare, primary health care services, prevention and control of diseases, etc., which form the main plank of our development efforts.

Broadly, health planning and programmes so far have worked to (a) control and eradicate communicable diseases, (b) provide preventive and curative health care system in the community development block and to augment the prevention of non-communicable diseases, (c) provide programmes to educate medical and paramedical personnel, (d) involve community level workers, that is, teachers, post masters, anganwadi workers, gram sevak and sevika in the delivery of health, (e) extend budgetary resources to voluntary organisations/institutions under grant-in-aid schemes.

The executive director of the Voluntary Health Association of India announced the formation of an Independent Commission of Health in India on January 24, 1995. The commission will assess the actual health situation in the country at the micro and macro levels. It will investigate areas like epidemics, public health institutions family planning and identify susceptible regions by dividing India into areas such as tribal and wild areas (for malaria), and the North-east (for AIDS and drug-related problems).

A number of national health programmes have been launched for the eradication of major communicable diseases. With the active cooperation of the WHO and State Governments in intensive compaigns, small pox has been completely eradicated. Progress has been made in controlling other diseases.

Malaria. The National Malaria Eradication Programme (NMEP) in the world's largest health programme against a single communicable disease and continues to be the country's most comprehensive and multifaceted public health activity. With the successful implementation of the programme in 1958, the annual incidence of malaria —caused by the malarial viral plasmodium and introduced by the anopheles mosquito into the bloodstream — was drastically reduced from 7.5 crore at the time of independence to about one lakh in 1965. Deaths due to malaria were completely eliminated. Unfortunately, due to various factors — unhygienic environment, DDT-immune variety of mosquito and the development of resistance of anti-malaria drugs by the parasites — these achievements could not be maintained. Resurgence of malaria necessitated renewed vigorous anti-malaria activities and the programme was modified in the context of escalating malaria incidence. The Modified Plan of Operation was implemented from 1 April 1977.

In India at present under the National Malaria Eradication Programme two projects (a) study on monitoring of sensitivity of P. falciparum to chloroquine (b) efficacy of alternate drugs of combination in chloroquine resistance strain of P. falciparum are being carried out. In-vitrotest on alternative drugs like Amodiaquine, Quinine Mefloquine and Sulpadoxin Pyrimethamin combination had also been conducted. Except Amodiaquine, all the other drugs were found sensitive. However the sample size is far below the norm.

Chemically treated nets that need to be treated every six months can keep the anopheles away. Nylon or polythylene nets are dipped into a solution of pyrethyroid, permethrin or deltamethrin in a plastic or aluminium container for about a minute and hung up to dry. The chemicals, whose dosage depends on the time the nets are allowed to drip, only kill the mosquitoes without having any effect on humans. The techniques can be used to treat curtains, etc. also. The nets thus treated prove useful even after they get torn. The nets can be used where DDT indoor sprays are no longer effective. They also eliminate other pests like ticks and bed bugs.

To combat the problem of emergence of malarial parasites that are resistant to the usual anti-malarial drugs, it is important to discover new drugs and rediscover old drugs and test them in new environments. The WHO has distributed low-cost in-vitro kits to doctors in Philippines to monitor parasitic response to anti-malaria drugs. The kits basically monitor the sensitivity of the Plasmodium falcipanum parasite to anti-malarial drugs in the field.

A malaria vaccine, SPF 66, developed by a Columbian has been donated by the Columbian Government to the WHO for worldwide distribution. It is the first chemically synthesised vaccine and the first vaccine against a parasite disease. It stimulates the immune system and attacks the parasites in the blood ; it thus prevents manifestation of the disease rather than the infection itself.

The Central Institute of Medicinal and Aromatic Plants (CIMAP), Lucknow, has been granted patent rights for production of artemisinin from the plant Artemista annua and converting the artemisinin into the anti-malarial drug 'Arteether'. The plant was introduced in the country by the CIMAP in 1986 and is being cultivated in the Kashmir valley. Artemisinin is active against chloroquine-sensitive and multi-drug-resistant strains of the parasite Plasmodium.

More recently, there has been an alarming rise in cerebral malaria caused by the parasite Plasmodium falciparum. Earlier limited to some north-eastern states and tribal areas, the disease is now visible in some northern states such as Rajasthan and the metropolitan cities too. A major factor leading to a growth in mosquito population is the environmental pollution consequent to uncontrolled industrialisation. A mobile population spreads the disease. The same parasite is also responsible for 40% of malaria cases. (The other parasite causing malaria is the Plasmodium vivax). The mosquito species Anopheles cultfacies is the major malaria spreading agent in the sub-continent. Canal irrigation in areas like Rajasthan has encouraged another major malaria-spreading mosquito, the Anopheles stephenst.

The Union Health Ministry in 1996 initiated a move to involve field-level government functionaries — at village, block, district levels — in the fight against malaria. The effort is directed towards preventing the creation of the conditions congenial to mosquito breeding; in other words community participation is to be mobilised to keep the environment clean.

The measures introduced by the NMEP include (i) increasing public awareness, (ii) conducting central level programmes on a rotational basis in states, Bhubaneshwar in Orissa witnessing the first such programme, (iii) greater attention to tribal areas, especially in-depth studies of the health care delivery system in these areas where there are cases of the deadly cerebral malaria or Plasmodium falciparum and poorly organised health facilities, (iv) reviewing drug distribution centres and fever treatment depots, (v) appointment of two experts in medical entomology to assist the Directorate, NMEP.

The disease is endemic in Bihar and West Bengal in India. The NMEP's Kala-azar unit is monitoring the disease in India. The control strategy envisages interruption of transmission for reducing vector population through indoor residual insecticidal spray, early diagnosis and complete cure of the disease, and building up community awareness. The Central provides insecticides and imported anti-Kala-azar drug. India's first depository for various strains of Kala-azar is coming up at the Indian Institute of Chemical Biology (IICB), Calcutta, recognised by WHO as a major centre for research into the disease. A special group at IICB was to employ modern techniques to understand the special features of the microorganism at the molecular level and develop new strategies for early diagnosis and development and delivery of drugs. Strains of Leishmania donovant bacterium causing Kala-azar are to be gathered and preserved. The will include strains resistant to existing antimonials—the most efficient drugs for the disease. The IICB has devised a new diagnostic technique that does not involve spleen or liver biopsy, but makes use of a DNA sequence found in the Kala-azar bacteria.

FILARIASIS. It is caused by a worm and transmitted by the mosquito. Nineteen States / Union Territories that include all the southern States while excluding most north-east States are endemic for filariasis. Estimates are that about 39.6 crore people live in 175 known endemic areas, about 2.5 crore have the parasites in their blood and more than a crore manifest the disease.

By blocking the lymphatic channels, the disease produces elephantiasis that effects 300-400 million people in the Indian subcontinent alone. In India, one of the worst-hit States is Tamil Nadu where around 10 per cent of the round about 35 million people living in areas endemic to the disease are infected. A three-decade-old idea of Frank Hawking, a British parasitologist, that of mixing salt with diethyl carbamazine citrate (DEC) has yielded a major mass treatment of this disease. DEC has no harmful side-effects and it is cheap and effective in small doses. It cuts down the chain of transmission of the disease. Field trials have indicated remarkable results—within 6-12 months, the infection rate declined from 13 to one per cent in Tamil Nadu.

For control of filariasis, the National Filaria Control Programme was launched in 1955. Its activities include delimitation of the problem in hitherto unsurveyed areas, and undertaking recurrent anti-larval and anti-parasitic measures to control the disease in urban areas.

Training in filariology is imparted at three Regional Centres—Calicut, Varanasi and Rajahmundry—under the National Institute of Communicable Diseases, Delhi.

Leprosy continues to be a major health and social problem in India. Out of 22 lakh leprosy cases in the world, nine lakh are to be found in India (1995 data), 15 to 20 per cent of these are infectious. All states report leprosy cases. The National Leprosy Control Programme started its operation in 1955 and in the year 1983, the programme was redesignated as National Leprosy Eradication Programme (NLEP) with the objective of arresting the disease in all the known leprosy patients by the year 2000 AD. It is a 100% Centrally-sponsored scheme. World Bank assistance has been taken for the programme.

Three candidate leprosy vaccines have been cleared by the Ministry for mass field trial to study their efficacy, which are being tried in the districts of Kolhapur of Maharashtra, Chinglepattu of Tamil Nadu and Kanpur in Uttar Pradesh. It would take at least five years to arrive at any conclusion regarding their effectiveness and utility under the programme.

About 7 lakh leprosy cases are receiving treatment. The treatment of leprosy cases with a number of drugs instead of one, has been found highly effective. Government has recognised the advantages of Multi-Drug Treatment (MDT) over traditional single drug therapy i.e. Dapsone monotherapy. All districts of the country are covered under MDT.

Several voluntary organisations are supplementing the governmental efforts in the fight against leprosy. A community-based comprehensive scheme has been drawn to rehabilitate leprosy-cured persons. In the first phase, it covers, on a pilot basis, five endemic districts—one each in Karnataka, Tamil Nadu, Andhra Pradesh, Maharasthra and Orissa.

Tuberculosis is another major health problem. About five lakh deaths occur every year due to tuberculosis. About fourteen million people in India are estimated to the suffering from radio-logically active tuberculosis of which 3 to 3.5 million are very infectious. The National Tuberculosis Control Programme has been given high priority. More than 18 lakh cases per year are being detected under the programme. At present district tuberculosis centres are functioning in many districts to undertake Tuberculosis Control Programme in association with general health and medical institutions. In addition to it, tuberculosis clinics are functioning mainly in urban areas. Primary health centres have also been involved actively in tuberculosis detection and treatment activities. In rural areas, multi-purpose health workers and village health guides are also being involved in motivation and health education activities under the programme. The National Tuberculosis Institute, Bangalore, established in 1959, is engaged in training medical and para-medical personnel and monitoring the National Tuberculosis Programme.

Essential activities under the programme have expanded considerably during the last few years. Duration of treatment of tuberculosis patients with conventional regimens is of 12 to 18 months and is believed to have been one of the major causes for irregularity and premature stoppage of the treatment by the patients. With the advent of Rifampicin and Pyrazinamide, it has become possible to reduce the duration of treatment of six to eight months. The National Tuberculosis Control Programme was revised in 1992 with assistance from the WHO and the Swedish Development Agency. Microscopic examination of sputum instead of X-rays is now the main detection tool. Short-term chemotherapy involving Rifampicin and Pyrazinamide and constant monitoring of patients' compliance towards the drug regime are components of the revised programme being introduced in a phased manner. So far over 250 districts have been brought under it. The Directly Observed Treatment Short-Course (DOTS), intended to monitor daily administration of drugs for the full term, has been tried out in the first phase of a pilot study in Mehsana (Gujarat), Calcutta, Bangalore and Delhi. The scheme relies on health and community workers.

Tuberculosis today is posing serious problems. The WHO Report in March, 1996 pointed out that multi-resistant tuberculosis, which often leads to death, has emerged as a global challenge. Tuberculosis is now the leading infectious disease killer of youth and adults. It has become the principal killer of HIV-positive people and is the major cause of maternal mortality. In India, a Planning Commission study says that the incidence of tuberculosis might be twice as high as the official claim. Cases of tuberculosis-IV co-infection are on the rise. The problem is mainly complicated by the patients not adhering to the full course of the drug regimen. Tuberculosis can usually be cured in a relatively inexpensive way by drugs like Rifampicin, Isontazid, Ethambutol and Pyrazinamide. These drugs often cure the patient of obvious symptoms such as blood-emitting cough and chest pain in about two months. However, these drugs have to be taken for at least another 4 to 6 months to kill all the infective microbes in the body. The average tuberculosis patient leaves off the drugs as soon as the symptomatic relief occurs ; as a result, some of the most resilient of the bacteria, which are still in the body, spring back into action soon enough. Now, multiple-drug-resistant (MDR) strains develop and "second-line drugs" (Ofloxicin, Cyclosporine) are required; these are more expensive, less effective and possibly toxic compared to the first-line drugs. Furthermore, poor patients are handicapped by not being able to afford the nourishing food required along with the drugs. The BCG vaccine has proved rather mediocre and only partly effective. It prevents the early, non-infectious forms of the disease in infants, but it can cause serious problems to patients who are HIV positive leading to what is called BCGitis.

More funds, more awareness and more dedicated health workers for adequate medical surveillance are required if the 'white plague', as tuberculosis is called, is to be tackled effectively.

Sexually transmitted diseases (STD) were placed under a national control programme as a centrally aided scheme in the Second Five Year Plan. It became a centrally sponsored scheme under the Fourth Plan. Now it has been merged with the AIDS Control Programme, as STD form a recognised factor in the transmission of HIV infection. The main emphasis in the control of STD has been on teaching, training, research, epidemiology and health and community education. Regional units have been established at Calcutta, Delhi, Hyderabad, Madras and Nagpur to function on zonal basis in the control of STD.

AIDS or Acquired Immuno Deficiency Syndrome has emerged as a frightening disease on a global level, and India has a fast-growing rate of AIDS patients. According to the figures provided by the

National AIDS Control Organisation (NACO) by May, 1995, out of about 25 lakh samples screened in India, confirmed HIV (Human Immuno-deficiency virus) infected cases was 18.291 and the number of AIDS cases was 1139. According to WHO, the highest percentages of adults (between 15 and 29 years of age) infected with HIV in Asia are in India.

AIDS was first detected in 1986 in India. Practically, all states now report HIV positive cases. The main cause of transmission of HIV in India is heterosexual activity; the second most common mode is transfusion of infected blood and blood products. In the north-east, Manipur has a large number of HIV positive cases mainly because of intravenous drug taking with infected needles. Manipur stands third in reporting HIV positive cases after Maharashtra and Tamil Nadu. Besides HIV-I strain, HIV-2 is also to be found in India. The National AIDS Control Programme was launched in 1989 with assistance from the World Bank and the WHO. The National AIDS Control Organisation was set up by the Union Ministry of Health and Family Welfare to implement the various activities under the programme. In August 1995, an AIDS Control Board was constituted in place of the technical advisory committee which gave direction to the national programme. The aim is to stop the spread of the infection and disease in India besides lowering the mortality of the infection in the country. The major components of the programme are as follows: (i) Information, communication and education to high risk groups as well as the general public; (ii) prevention and treatment of STD; (iii) prevention of transmission through intravenous drug abusers; (iv) prevention of transmission through blood and blood products; (v) strengthening clinical management capabilities; (vi) programme management.

Under the Drugs and Cosmetics Act, selling human blood contaminated by HIV or hepatitis B-virus is punishable with life imprisonment.

There are many surveillance centres for screening persons in the high risk groups. In addition, zonal blood testing centres have been established in metropolitan cities for screening blood donors.

Drug de-addiction centres have been provided in major hospitals. To contain the spread of AIDS among intravenous drug users in Manipur, the Central Government has approved Needle Exchange Programme (NEP) for the state. The programme was launched in January, 1996. Under the programme, non-government organisations can distribute needles to drug users and open out-reach centres from where users can obtain fresh needles, free of cost, instead of sharing them. Manipur is the first Indian State where such a programme has been launched.

The Human Immuno-deficiency Virus (HIV) causes the dreaded AIDS— Aquired immuno-deficiency syndrome. As to the origin of the HIV, it is postulated by a majority of researchers that it originated in Africa from a wide range of sub-human primates within the last 50 to 100 years. Other theories: it could have existed for generations in isolated human communities and spread with the increased mobility of these communities; it may have resulted because of a mutation in the wake of mutation in the wake of radioactive fallout. The HIV is a retrovirus, genetically highly variable in antigenicity because it lacks the remediation intrinsic to DNA replication. It integrates into the host cell for the life of the host and produces an effective cytotoxic T-cell response. The major targets of the virus are the cell populations responsible for initiating and maintaining an immune response — the lymphocyte. While HIV-1 is common worldwise, HIV-2 is also spreading, through it is mainly to be found in Western Africa. In India HIV-1 and HIV-2 infections exist.

An accurate count of CD-positive helper T lymphocytes is required to diagnose AIDS which is considered to exist if there are less than 200 CD, per cubic millimetre.

HV is transmitted through blood and blood products, seminal and vaginal fluids, during the following:-

— Unprotected sex (without a condom), especially anal, with an infected person;

— Receiving infected blood in transfusions, blood products, organ transplants, through contaminated needles, syringes or other skin-plercing instruments;

— From an infected mother to a baby before, during or after birth via the placenta and through breast milk;

— Through artificial insemination with infected semen.

— HIV is not transmitted by:

* hugging, kissing, shaking hands.

* sharing cups and vessels.

* coughing and sneezing.

* mosquito or insect bites.

* food and water.

Signs. The major signs are:

— Weight loss greater than 10 per cent of body weight in one month.

— Chronic diarrhoea for over one month, either intermittent or constant.

— Prolonged fever for more than a month, either intermittent or constant.

Minor signs are:

— Persistent cough for more than one month

— Generalised swollen lymph nodes

— Herpes Zoster infection

— Persistent night sweats

— All homosexuals who have been sexually active in the last few years.

— All haemophiliacs who have received Factor VIII.

— Everyone who has had a transfusion of blood or blood products (plasma, platelets, etc.) In the last few years.

— Individuals who have had sex (especially anal sex) with a prostitute.

— All intravenous drug users who have ever shared a needle or syringe.

— All sexual partners of intravenous drug users.

— All children born to HIV-infected mothers.

— All donors of sperm, blood, organs for transplantation.

The main laboratory tests to detect the virus are the ELISA (Enzyme Linked Immunosorbent Assay) test, Western Blot test which is highly specific but requires trained and experienced workers to interpret, and the virus isolation method which is effective but expensive. Other tests are particles agglutination test (PAT), immuno-flourescent assay (IFA), radio-immuno precipitation assay (RIPA). In India, two simple and reliable diagnostic kits to detect the virus in blood and blood products have been made available. The Combaid Visual dip-stick test technology was obtained by Span Diagnostics Ltd. Surat and the kit is manufactured by them. The Hiva Test is the first and only indigenously developed HIV-detection kit; it was designed by the International Centre for Genetic Engineering and Bio-technology at New Delhi. Both Kits are immuno assays based on the specific binding ability of antibody molecules able to detect the presence of antibodies against HIV-I as well as HIV-2.

While AIDS is almost always a fatal disease, some drugs have come up which are useful in treating AIDS. Azidothymidine (AZI) is one such product: the AZT molecule interferes with the process by which the AIDS retrovirus infects the cell and replicates itself. Zidovudine is the name of the AZT drug produced by the Indian Institute of Chemical Technology, Hyderabad. Dideoxyinosine (DDI) is another drug which is useful for AZT-intolerant patients. Zalcitabine is effective in the early stage of the virus replication by inhibiting reverse transcriptase and ending the DNA chain.

A new drug, Saquinavir, is the world's first rationally designed antiviral drug. After seven years of work, it has been submitted for a licence in the USA and Europe. It attacks the virus responsible for AIDS, HIV in a new way. It inhibits the enzyme called protinase, which helps the virus mature and spread, by cleaning the long-chain where the structural proteins needed to build new viral particles are generated.

As for vaccines, the initial euphoria has abated, as the various sub-types of the HIV-I make the development of an effective vaccine rather difficult.

India has perhaps the largest blind and potentially blind population in the world. The major cause is cataract (about 80%)

followed by trachoma, small pox, malnutrition/Vitamin A deficiency, injuries and glaucoma.

Recognising that 85 per cent of blindness is either preventable or curable through treatment by modern medical/surgical intervention, a comprehensively conceived National Programme for Prevention of Visual Impairment and Control of Blindness now known as the National Programme for Control of Blindness (NPCB) was launched by Government of India in 1976. It aimed at reducing the prevalence of blindness from 1.4 per cent to 0.3 per cent by the end of the century. The four-fold approach of this programme was to (i) intensify educational efforts on eye health care, (ii) provide immediate relief to people with cataract and other eye diseases in far flung areas through mobile units and camp approach, (iii) establish permanent properly-equipped eye care infrastructure at various levels, including development of ophthalmic and ancilliary manpower and (iv) involve voluntary/non-governmental organisations and private sector in eye health care.

Cataract is curable. Most of the cataract operations are performed by voluntary and non-governmental organisations. The need is to step up cataract surgeries to four million or more annually from 1.2 million annually. Heavy backlong of cataract cases can be addressed by (i) optimising the use of the existing ophthalmic manpower and hospital facilities, (ii) inducting management strategies, (iii) encouraging new, innovative approaches, (iv) establishing many base eye hospitals especially at the sub-divisional levels.

A Rs.554 crore, World Bank assisted, seven year project for checking cataract blindness was launched in October 1994 in Tamil Nadu, Andhra Pradesh, Maharashtra, Uttar Pradesh, Orissa and Rajasthan States which account for nearly two-thirds of the country's blind population and one-fourth of the world's. The project will improve the quality of cataract survey and reduce the huge backlog of cataract cases. The efforts would include introduction of cost-effective systems of managing hospital and eye camp-based operation and maintenance of high standards, strengthening the States' technological capacity and developing institutional capacity and evolving appropriate coordination mechanisms.

Nearly 35,000 children go blind annually owing to lack of Vitamin A. Vitamin A prophylaxis for the children is being conducted nationwide since 1971 as part of the Maternal and Child Health (MCH) Programme of the Department of Family Welfare of Government of India. Its coverage through MCH extends to about 60 per cent of the eligible children. A recent survey indicates that prevalence of nutritional blindness due to Vitamin A deficiency has come down to 0.04 per cent from 2 per cent two decades ago. There should be a campaign for the consumption of green leafy and yellow coloured vegetables and fruits. They are easily available at low cost and are rich in Vitamin A.

Visual impairment due to refractive errors amongst the children has assumed special prominence. This is being combated by intensifying eye screening of the children by MPCB's mobile units and various voluntary organisations functioning centrally and through its state and district branches. An innovation introduced is involving the rural school teachers in the screening programme.

A considerable number of people lose their eye-sight due to accidents which can be prevented. Eye injuries through crackers and fireworks occur during the festivals. In welding industries, stone breaking units etc. the protective gadgets that are suggested by the industries are ignored, the shop floor level instructions are not displayed, many employers do not provide the eye shields etc. Even when these are supplied they are not used by the workers should be educated in the value of such protective gadgets and their proper use.

Sustained education of the workers, industrialists, school children, unorganised sector units and suitable statutory measures to ensure use of the safety methods should be undertaken. This calls for involving all mass media channels, electronic and others, and optimising the use of existing extension education network.

The human body requires 150 micrograms of iodine daily. Ninety per cent of this requirement is normally taken care of by the food and water intake by a healthy individual. Iodine content is drawn from the soil by the food grown in it and water that flows through it.

Iodine being a water soluble element is generally present in the superficial layers of the soil. Due to various reasons like erosion of soil by snow, rain, floods and rivers changing their course, large amounts of this previous element is carried away by rivers into the sea. With the passage of time the iodine content of the soil is getting depleted.

The thyroid gland uses iodine to synthesize thyroxine which helps in growth and development. Iodine deficiency results in less of thyroxine and thus reduced blood circulating levels. Earlier, goitre was seen as the only indication of lack of iodine. But a number of disorders can be related to iodine, such as spontaneous abortions, both of cretin children, neonatal hypothyroidism and as much as 10 per cent drop in the intelligent quotient in children.

Iodine deficiency is one of the major widespread and easily preventable nutritional deficiencies prevailing in the developing world.

In India, it was earlier believed that the endemic goitre belt stretched across sub-Himalayan regions. With this background, National Goitre Control Programme (NPGC) was launched in 1962 with 100 per cent assistance from the Central Government with the following objectives : (i) to identify the goitre endemic regions, (ii) to supply iodized salt in place of ordinary common salt to goitre endemic areas, (iii) to assess the impact of goitre control measures over a period of time.

The NGCP has now been designated as National Iodine Deficiency Disorders Control Programme.

As surveys progressed, additional areas with goitre-affected population were identified. Till date, sample surveys conducted in 243 districts covering all the 25 States and four out of seven UTs revealed that 200 districts are endemic for iodine deficiency disorders (IDDs). In India 6.3 crore people are estimated to be suffering from various IDDs. It is not poverty alone but malnutrition as such that forms a common ground for iodine deficiency in the country. In an effort to curb the problem, the Central Council of Health, Government of India started the "Universal Iodization of Entire Edible Salt in the country by 1992" scheme, to be implemented in a phased manner, in April, 1986. Common salt has been chosen for iodine supplementation owing to its consumption patterns, limited production sources, easy operation of quality control measures with regard to it, its cost effectiveness and easy acceptability and no change in palatability where it is concerned.

As potassium iodate is stable even in warm, humid climate, it has been selected as an iodinating agent. The country's vast size and its transportation problems have also been taken into consideration.

It is proposed to extend the legal provisions stipulating a minium of 30 ppm of iodine at the manufacturing level and 15 ppm on wholesale/retail level — till date effective in about 60% areas — to cover the entire country. A legal ban on sale of non-iodised salt has been imposed. It has been made mandatory for all manufacturing units to have quality control laboratories at manufacturing points. Iodised salt production has been opened to the private sector. Hundred per cent financial assistance is also provided to states and union territories for establishing a goitre control cell to ensure proper implementation and monitoring of the goitre control programme. The Directorate-General of Health Services who is responsible for implementation of the programme at national level will also take over monitoring and evaluation work. The district hospitals of endemic districts and selected medical colleges will undertake iodine intake monitoring through urinary iodine excretion. Portable testing kits have been made available to ensure quality of iodised salt.

Allegations have been made regarding iodine abuse and iodine toxicity. The rare complications include allergic reactions to iodine in food, iodism and iodide goitre, iodine induced hyperthyroidism. The toxic effects of iodine are observed only if large pharmaco-logical dose of iodine are taken. Taking into account the level at which potassium iodate is being used for iodating salt, an average individual with a body weight of 60 kg will have to consume 300 kg of salt in a day which would contain 30 gm. of potassium iodate to die of potassium iodate toxicity. The study group of World Health Organisation in its final report has opined that there is no danger of toxic effects if todates are substituted for iodides in salt iodation.

CANCER. Cancer is a disease which carries a high rate of mortality unless it is detected and treated early. In India lakh cases and about six lakh new cases occur every year. The mortality rate is very high.

Fluorosis is caused by fluorine, a normal constituent of natural water that forms a higher content in groundwater than in lakes and streams, and still less in deep-well waters. Long-term ingestion of water containing 8 ppm or more of fluoride causes skeletal fluorosis in man as fluoride accumulates in bones as fluorapatite. Fluoride causes disturbances in ameloblasts and reduces the cementing substance, causing mottling of teeth. Bone density increases, bone substance is abnormally deposited and ligaments and joints become hard and immovable due to calcification. Factors that influence the severity of skeletal fluorosis are total daily intake, concentration, age, nutritional status, individual biological response, climatic factors, pH of gastrointestinal tract.

Poor nutrition, calcium deficiency and hard manual labour have contributed to an increasing incidence of fluorosis in India. The 'genu valgum' form of fluorosis prevalent in India is characterised by crippling knock knees, hardening of the soft knee joints and wasting of bone tissue in the limbs. In India fluorosis is reported from areas where the fluoride intake is less than the permissible level of 1 ppm. It is estimated that about 25-30 million Indians in 150 districts are suffering from varying grades of fluorosis. Studies by the Rajiv Gandhi National Drinking Water Mission have revealed that of the 15 states identified as endemic of fluorosis, Uttar Pradesh, Rajasthan, Gujarat, Andhra Pradesh and Tamil Nadu have 50-100 per cent of the districts affected while Punjab, Haryana, Madhya Pradesh, Maharashtra and Karnataka have 30 to 50 per cent of the districts affected. The highest levels, at 38.5 ppm, have been found in some villages in Gurgaon, Haryana. In some villages, nearly 75 per cent of the population has been affected. Studies have also disproved the belief that safe and contaminated sources do not co-exist. This means that each sources has to be tested individually; checking a few samples can be grossly misleading. Analysis of fluoride content has, however, become accurate with the gifting of 45 ion analysers by the WHO. The analysers have been distributed to district level laboratories in states.

As there is no cure for fluorosis, ingestion of fluoride must be kept at low levels. Use of water from deep bore wells helps in eradicating and containing fluorosis. Defluori-dation of water can be done by chemical treatment of water. Special filters for the purpose have been developed by the Sri Sat Sai University for higher learning at Anantapur, AP. The National Environment Engineering Research Institute (NEERI) has developed a technique involving rapid mixing of water with lime, alum and bleaching powder. A control on tobacco consumption and alcohol consumption will greatly help as these increase acidity in the gastrointestinal tract and increase fluoride intake. Steps to check indiscriminate digging and spread awareness of the importance of quality of water are needed. Fluoride content in tooth-paste should be checked.

Dracunculiasis or guinea-worm disease, would be wiped out from the face of the earth within the next few years. The World Health Organisation has reported a decline in the disease; complete eradication should be achieved on target by 1995, according to WHO. It will be only the second disease to be utterly defeated after small-pox, wiped out in 1979 as a result of WHO programmes and massive voluntary support provided from all over the world.

Guineaworm infection is contracted by drinking water infested with cyclops, a minute crustacean predator that forms part of the plankton of inland water. The cyclops swallows one of millions of larvae released into water by the female guinea worm. After two weeks, the larvae become infective. When humans drink the water, gastric juices dissolve the cyclops, freeing the larvae, which then grow and mate within the human body.

A year after a person has swallowed the cyclops, the mature female worms force their way through soft tissue of the body causing severe pain and blistering on the skin before finally breaking through the surface, usually on the ankle or foot. The blisters on the skin often break when the foot is immersed in water, and as they do so, the emerging worms gradually release hundreds of thousands of larvae. They then swim off to be swallowed by the cyclops, renewing the cycle of transmission.

There is no drug treatment for the disease. The worms up to one metre long, can be extracted. But healing of wounds they cause offers no immunity from further infection. Despite these problems, the disease is being overcome because of successful public health education programmes, including basic advice on the filtration of drinking water. Piped water supplies and concrete lined wells have been provided, as have insecticides to treat water supplies and specially-designed nylon cloth filters for families. States suffering from this disease are Rajasthan, Andhra Pradesh, Gujarat, Karnataka, Madhya Pradesh and Maharashtra. India expects to be free of the disease by the end of this century.

After the occurrence of Japanese encephalitis (JE) epidemic in different states in 1977-78, the Government reached an agreement with the Government of Japan for the manufacture of Japanese encephalitis vaccines. This vaccine is being manufactured at CRI, Kasauli, Maximum cases are reported from Tamil Nadu, Bihar, Assam, Andhra Pradesh, Uttar Pradesh and West Bengal, JE control activities are being implemented in problem a states under NMEP.

The Government of India decided to launch the National Mental Health Programme during 7th Five Year Plan period with the objectives to ensure availability and accessibility of minimum mental health care for all, in the foreseeable future, particularly to the most vulnerable and under-privileged sections of population; to encourage application of mental health knowledge in general health care and in social development; and to promote community participation in the mental health service development and to stimulate efforts towards self-help in the community. A National Advisory Group on Mental Health has been constituted under the chairmanship of Secretary. Ministry of Health and Family Welfare for the effective implementation of the National Mental Health Programme.

The National Diabetes Control Programme was included in the 7th Five Year Plan as one of the Central Health Sector Programmes. It was recommended that infrastructure for monitoring and evaluation at the national level be created during 1986-87 with identification of districts and initiation of exploratory contacts with the State-level health functionaries so as to begin implementing the programme. The District Diabetes Control and Care Programme was launched in two districts of Tamil Nadu, namely Salem and South Arcot, in line with the National Programme. The objectives are (a) Identification of high risk subjects at early stages and imparting appropriate health education with focus on primary prevention of diabetes, (b) Early diagnosis of disease and institution of appropriate management so as to reduce morbidity and mortality (secondary prevention) with emphasis on volunerable groups, e.g. gestitional diabetes, (c) Prevention, arrest or slowing of acute metabolic as well as chronic cardio-vascular-renal complications of the disease, (d) Provision of equal opportunities to ensure holistic as well as physical attainment and job satisfaction thus ensuring social and emotional adoption leading to an improved quality of life, (e) identification of those with partial or total physical handicaps dure to disease to ensure their rehabilitation with emphasis on optimal organ or body function.

It is also aimed to prove highly specialised units capable of providing laboratory as well as management skills for the care of the diabetic. Another dimension emphasises the research component. A beginning in this direction has already been made by the research section of Diabetes Association of India and various expert groups have been constituted.

Recent studies on the cause, occurrence and control of diabetes have emerged with some findings that promise to have a major impact on diabetes management. Of all the peoples in the world, it is said that the disease has its greatest prevalence among Indians. It has been reported from the UK, South Africa, Singapore, Fiji and Trinidad that Indians living in these areas are more prone to the disease than their own peoples. Five to 10 per cent of the urban population in India suffers from diabetes. To major studies, the landmark 10-year Diabetes Control and Complications Trial done on insulin dependent diabetes mellitus (IDDM) in the US and the Prospective Diabetes Study done on non-insulin dependent diabetes mellitus (NIDDM) in the UK, have conclusively proved that the risk of complications could be reduced by 60 per cent with intensive treatment and tight control.

There is fresh evidence that people are genetically prone to complications. In the above mentioned studies, it was found that even among the tight control groups some people did develop complications. In the non-intensively controlled group, however, some people did not develop complications. Studies at the M.V. Diabetes Specialities Centre in Madras have shown a strong possibility of a genetic link. In the case of non-insulin dependent diabetics with proliferative retinopathy, an association has been noticed with the immunoglobulin switch gene on chromosome 14. This has been supported by the findings of a group in Britain. These findings show that surveillance should be started early for diabetes and complications, if detected at an early stage, are treatable or potentially reversible.

Fibrocalculous Pancreatic Diabetes (FCPD). A disease where stones in the pancreas lead to inflammation and diabetes, FCPD is a disease of developing and tropical regions, with the highest prevalence in the world in South India. It was first discovered in 1958 in Indonesia. Studies have shown that it could be a disease related to micronutrients deficiency as Vitamin C and beta carotene levels have been found to be low in FCPD patients. The patients may develop complications including pancreatic cancer. Genetic work on this disease has shown overlap in genetic susceptibility between IDDM and NIDDM.

CONTROL. Acute diarrhoeal disease is one of the major causes of morbidity and mortality in India, especially among infants and children below 5 years of age. However, no reliable data on diarrhoea-related morbidity and mortality are available as diarrhoeal diseases (except cholera) are not notifiable and over 25 different pathogens may cause the disease. However, significant strides made in the last decade on the etiology, epidemiology, pathogenosis, immunology and treatment of acute diarrhoeas have helped in addressing the problem.

The World Health Assembly in May, 1978 called for a concerted attack on the diarrhoeal diseases as part of the global commitment to primary health care and to 'Health for All' by the year 2000 AD. In the light of the recommendation the National Diarrhoeal Disease Control Programme was launched in 1981. In order to attain the objective of reducing childhood mortality and the long-term objective of reducing morbidity caused by the diarrhoeal disease the following approaches are being promoted : (i) to reduce mortality with Oral Rehydration Therapy as early as possible in the course of illness and educating mothers for appropriate feeding of children, (ii) To reduce morbidity through (a) the encouragement of maternal and child care practices such as uninterrupted breast feeding, preparation of safe weaning food, good domestic and personal hygiene and adequate nutritional support for pregnant and lactating mothers, and (b) better environmental health practices especially the proper use and maintenance of drinking water and sanitation.

Oral Rehydration Salts packets, effective for all age groups, are available throughout the country also through commercial pharmaceutical outlets. The top priority component of the programme is training of the health personnel at all levels. 90% of diarrhoea cases have no dehydration. Dehydration can be prevented by use of butter milk, rice water, green coconut water etc.

On April 21, 1995 a National Plan of Action for Nutrition was released to emphasise the methods of Central and State Government sectors to attain the aims of the earlier-formed National Nutrition Policy. The Plan, prepared by the Department of Woman and Child Development of the HRD Ministry, would evolve strategies to fight malnutrition and improve the health of women and children. A common action front involving the departments of Agriculture, Health, Food and Civil Supplies, etc. would be evolved.

Polio

Polio is a dreadful communicable disease of especially developing countries. Two out of three cases of polio in the world are recorded in India.

Polio mainly targets children below five years of age. The polio virus is generally believed to spread from the nose, throat and intestines — that is, through sneezing, coughing, etc. The viruses that cause polio are called types I, II and III. Polio mainly spreads in areas of poor sanitation. The virus can survive in sewage for as long as three months. Contaminated food and water is a major mode of transmission. Entering the body through the mouth, the virus multiplies in the intestines. All people infected by the virus do not get polio. Some may show mild symptoms like fever, headache, sore throat, vomiting that might disappear after a day or recur and lead to severe polio. In about one out of every 200 cases, the virus passes to the spinal cord and affects the nerve cells that activate the muscles to cause lifelong paralysis. When the nerve cells in the brain get affected, respiratory failure and finally death results. In children below three it can prove fatal but this is rare. Fatality rates are between 10-15 per cent, half of which may result in paralysis.

There is no drug to kill the virus or control the spread of polio. There are two vaccines — the Oral Polio Vaccine (OPV), a live vaccine developed in 1961 by Dr. Albert Sabin and taken by mouth, and the injectible killed or inactive polio vaccine introduced in 1955 by Dr. Jonas Salk.

Polio Prevention. To encourage efforts on polio eradication, the WHO has declared the year 1995 as dedicated to polio eradication. The aim is to eradicate polio by 2000, that is, TARGET 2000-A World Without Polio. The WHO's strategies for polio eradication include (i) support routine immunisation systems (ii) conducting National Immunisation Days (NIDs), (iii) establishing sensitive surveillance systems and use of surveillance data to plan and conduct localised immunisation campaigns where OPV can be taken to the doors.

Six emerging polio-free zones — the Americas, Western and Central Europe, North Africa, West Asia, Southern and East Africa, and the Western Pacific — are places from where the wild polio virus has either been eradicated or where it is at a very low level.

Efforts in India. The aim in India is to have O cases of polio by 1997, and eradication of the wild polio virus itself by AD 2000. There has been a decline of polio cases over the years — 4003 in 1994 as against 4,500 in 1993 and 24,257 in 1988. Delhi has the highest incidence. For polio prevention, as many as 14 lakh children were immunised in October 1994. The country's first polio-free zone is Pattanamthilta district in Kerala. OPV, introduced in the National Immunisation Programme (NIP) in 1987, is seen as the best answer for mass immunisation.

A major thrust for polio prevention is the recently introduced NIDs programme of the WHO that employs the pulse immunisation technique — large numbers of a susceptible population receive OPV in one or two days. It has proved highly successful in other countries like Brazil and China. The Pulse Polio Programme has proved a success in Vellore, Madras and to a certain extent in Bombay and Delhi also. Plans are to extend the programme so that all children below the age of three can be immunised on two specific days of the year all over the country. Rupees 50 crore has been sanctioned for the programme which will be carried out over four years.

The Indian Council of Medical Research is the apex body in India for the formulation and promotion of biomedical research. The primary functions of the Council are discharged through its permanent research institutes / centres (20 at present), regional medical research centres, centres for advanced research, research units, task force projects and national multicentric collaborative projects in different parts of the country. In addition, there are ad hoc schemes and fellowships generated by active scientists in biomedical institutions / universities in different parts of the country. ICMR also attempts to strengthen indigenous capabilities, promote a broad-based and balanced cadre of research personnel in the country and develop facilities to tackle the present and future health problems.

The Council's research priorities coincide with the national health priorities such as control and management of communicable diseases, fertility control, maternal and child health, control of nutritional disorders, developing alternative strategies for health-care delivery; containment within safety limits of environmental (and occupational) health problems; research on major non-communicable diseases like cancer, cardiovascular diseases, blindness, diabetes and other metabolic and have matological disorders, and mental health research and drug research (including traditional medicine). All these efforts are undertaken with a view to reducing the total burden of diseases and to promote health and well-being of the people. (India 1996).

Research in traditional systems of medicine has adopted a disease-oriented approach. Biomedical informatics has been intensified to bridge the existing gaps between researchers, health planners, policy makers, providers and users of health services. Apart from the print media, the audio-visual media is being used more and more for dissemination of biomedical information.

There are two statutory bodies to conduct research in various branches of medical sciences. These are : All India Institute of Medical Sciences, New Delhi and Post-graduate Institute of Medical Education and Research, Chandigarh. There is also an All India Institute of Speech and Hearing at Mysore. Clinical facilities are available at all these Institutes.

The Central Research Institute, Kasauli is the biggest and most important producer of immunobiologicals, viz. DPT, anti-typhoid, anti-rabies, yellow fever vaccine and cholera vaccine and also for anti-sera and diagnostic reagents antigens. This is the only Institute producing yellow fever vaccine in South East Asia region. The Institute trains candidates sponsored by WHO, other international agencies and vaccine production organisations in the country in the techniques of production of various vaccines and sera and their quality control. The Pasteur Institute of India, Coonoor, Nilgiris is engaged in conducting research in rabies, influenza, other respiratory virus infections, etc. and in the other respiratory virus infections, etc. and in the production of anti-rabies vaccine and other vaccines.

A broad based infrastructure has been created for taking primary health care services to the people nearest to their homes and at a cost within their reach. Over the years, emphasis has been shifted from hospital based urban care to field based rural health care. Mother and child health care services are a vital part of primary health care. The targets for a sub-centre for 5000 (3000 in hilly and tribal areas) population and primary health centre for 30,000 (20,000 in hilly and tribal areas) population have been almost fully achieved. At present 1,31,476 sub-centres, 21,254 primary health centres and 2,328 community health centres are in action as a part of huge network to mitigate the health problems and uplift health standards. Of the people residing in the villages, special attention is being given to the interior places of the rural areas where so far the health care benefits have not reached. What is required today is to make this infrastructure function more efficiently and effectively.

Immunization is a low cost efficient strategy for children's health. An expanded programme on immunisation (EPI) was launched in 1978. It was given further impetus when an ambitious universal immunization programme (UIP) was started in November, 1985. The programme aims at providing protection to all children against six vaccine preventable diseases — Neo-natal tetanus, Diptheria, Pertussis (whooping cough), Poliomyelitis, Tuberculosis (childhood) and Measles. All expectant mothers are to be given tetanus toxoid injections to protect both the mother and the child from tetanus. The programme has covered the country's 452 districts.

Before 1985, immunization services were provided under the family welfare programme, but the scheme was limited primarily to the major hospitals. Coverage level was, therefore, very low. Since then, however, the pace has been stepped up. When the programme began in 1985, the magnitude of the task was stupendous. Infant mortality was then at a relatively high level of 97 for every 1000 live births, and it was estimated that vaccine-preventable deaths were responsible for one-fourth of all infant deaths. Nearly 3,00,000 infants died of neo-natal tetanus every year, while another 150,000 were vulnerable to paralytic polio.

Today, the programme has achieved 100 per cent success to cover all infants and pregnant women. Every year, about 22 million infants are vaccinated with three doses of the DPT, three doses of the polio vaccine, and one dose each of the BCG and measles vaccines. Besides, 25 million pregnant women are also administered two doses of tetanus toxoid every year.

Infant mortality has fallen substantially from 97 per 1000 live births in 1985 to 74 in 1993. In some states like Tamil Nadu, Maharashtra and Punjab, the figure is below 60, and in Kerala it is as low as 13. The Hindi belt states of Uttar Pradesh, Rajasthan, Madhya Pradesh and Bihar are the laggards, with Orissa having a high infant mortality of 110 per 1000 live births.

The incidence of most vaccine-preventable diseases has also declined impressively. The immunization mission is now considering the possibility of declaring Haryana, Himachal Pradesh, Karnataka, Maharashtra, Punjab, Tamil Nadu, Goa, Chandigarh and Pondicherry "polio-free zones" within two years. A similar decline has also been reported in the incidence of diphtheria and pertussis.

One other fallout of the immunization mission is that self-sufficiency has been achieved in the manufacture of all vaccines except the polio vaccine and BCG. When the programme began in 1985, the measles vaccine was imported entirely. Today it is being manufactured indigenously at the Serum Institute of India, and a surplus capacity of 25 million doses has even been built up, so that exports are a distant possibility.

On the anvil are plans to begin the indigenous production of the polio vaccine, and to double the indigenous capacity of BCG from 25 to 50 million doses.

While efforts are on to provide literacy and immunity from disease to semi-urban and rural populations, yet another mission that is racing against time is the one to provide safe drinking water to all villages in the country.

At the beginning of the Seventh Plan (April 1985 there were 1.62 lakh "problem villages" in India, where safe drinking water was not available within a 1.6 kilometres radius, underground water sources were deeper than 50 metres, and the water sources were chemically and bacteriologically infected. The drinking water mission was launched in August 1986. The mission has since been renamed as Rajiv Gandhi National Drinking Water Mission. By the end of the Seventh Plan (March 1990), 8,365 problem villages remained to be covered. According to data available, 750 of these villages remained to be covered in 1993-94. About 78 per cent of rural population and about 84 per cent of urban population have been supplied with drinking water.

In the Eighth Plan augmentation of safe drinking water supply in 1.5 lakh partially covered villages is contemplated. New no source habitations are emerging due to drought or depletion of water table. Special attention is to be paid to coverage of scheduled caste/ scheduled tribe population and other weaker sections under the programme.

Economic and social progress are both adversely affected by high population growth. The 1991 Census puts our country's population at 846.3 million, showing a 23.9 per cent growth in the decade from 1981. The Eighth Plan aims at reducing the crude birth rate from 30.2 per thousand in 1990 (estimated to be 29.0 per thousand in 1992) to 26 per thousand by 1997. Infant mortality rate is also to be brought down from 80 per thousand live births in 1990 to 70 per thousand live births by 1997. The Population Policy of 1983 and the Seventh Five Years Plan set a target of achieving a Net Reproduction Rate (NRR) of unity by the turn of this century; however, the Eighth Plan document has shifted the target to the period 2011-2016 AD. While socio-economic factors must guide family welfare plans and programmes, science and technology has an importance role in this sector, both in devising ways and methods of controlling population growth and in ensuring good health for mother and child to reduce maternal and infant mortality.

In the field of contraceptives and other birth control methods, science and technology has made available the condom, the pill and surgical intervention like sterilisation. The Government has made efforts to provide contraceptives such as Nirodh free of charge through the public distribution system.

The programme of medical termination of pregnancy supplements the Family Welfare Programme as it provides for legalised abortion in cases of contraceptive failures. A good proportion of the acceptors of abortion go in for some form of contraception like sterilisation, intra-uterine device (IUD), etc. A colour test 'Pregelisa' (Pregnancy Test Kit) has been developed. The kit is under rigorous testing procedures and if found suitable, will be marketed throughout the country including rural areas.

Now, there is the injectible contraceptive for women, Depo Provera, which is being introduced in the country, though controversy surrounds its side-effects. Depo Provera is the brand name for Depot Medroxy Progesterone Acetate (DMPA). It is a hormonal contraceptive which causes an increase in the level of progesterone in the body. As a consequence, the brain suppresses the release of the gonadotropic hormones which stimulate the maturation and release of the mature egg.

Scientists at the Indian Institute of Chemical Technology (IICT) in Hyderabad have successfully produced the abortion pill RU-486, thereby breaking the French monopoly. In combination with prostaglandin RU-486 works as a safe abortificient for use post-coitally.

The safe motherhood and child survival programme plays the most important part in family welfare programmes. The programme ensures effective ante-natal care, safe and asceptic delivery, care of mother and infants for immunization, control of diarrhoeal diseases and acute respiratory infections, nutrition education and provision of basic medical care.

Every ancient civilization developed systems of curing and preventing disease. All of them make use of plant products in their medicines. In recent times Indian systems of medicine have been getting more and more prominence as these, unlike allopathic drugs (especially antibiotics), do not produce drastic side-effects.

Indian systems of medicine (ISM) include Ayurveda, Siddha, Unani, Amchi, Tibetan and Naturopathy. Homoeopathy, though of German origin, is quite popular in India. Several institutions have been established to develop these systems in India, and there are about four lakh practitioners of ISM today.

The Central Council of Indian Medicines and the Central Council of Homoeopathy are responsible for maintaining uniform standards of education and regulating professional practice in these fields.

Four research councils viz. (i) Central Council for Research in Ayurveda and Siddha (CCRAS), (ii) Central Council for Research in Unani Medicine (CCRUM), (iii) Central Council for Research in Homoeopathy (CCRH) and (iv) Central Council for Research in Yoga and Naturopathy (CCRYN) are registered societies fully financed by Government.

Government has established a public sector undertaking viz. Indian Medicines Pharmaceutical Corporation Ltd. with the primary objective of producing quality Ayurvedic, Unani and Siddha medicines. The factory and the registered office are situated at Mohan, Almora District.

Pharmacopoeial Laboratory of Indian Medicine and Homoeopathic Pharmacopoeial Laboratory, both at Ghaziabad, are subordinate offices of the Ministry of Health. The former is engaged in laying down the standards of single drugs and compound formulations of Ayurvedic, Unani and Siddha systems of medicine, while the latter functions as a standard-setting-cum-drug testing laboratory at national level. This laboratory also works out various standards.

The Rashtriya Ayurveda Vidyapeeth, New Delhi, National Institute of Ayurveda, Jaipur, National Institute of Homoeopathy, Calcutta, the National Institute of Unani Medicine, Bangalore, National Institute of Naturopathy, Pune are working in various ways to promote ISM and encourage research in these fields.

The Indian Medicinal Plants National Network of Distribution of Databases has been launched to boost local health traditions in the country. The project is being serviced by the Foundation for Revitalisation of Local Health Traditions, a non-government organisation set up in 1991 and comprising eminent scientists,- medical experts, community health organizations and Government officials.

The nine nodal agencies under Inmed plan will manage individual databases in a specialised field. They include the Central Institute of Medicinal and Aromatic Plants, Lucknow, for agrotechnology; Botanical Survey of India, Calcutta, for botany; French Institute of Pondicherry, for ecology; Tropical Botanic Garden and Research Institute, Kerala for pharmacology; Central Drug Research Institute, Lucknow. The others are the Regional Research Laboratory, Jammu, for phytochemistry; Centre for Indian Medical Heritage,

Coimbatore, for traditional systems ; and the Publication Information Directorate, New Delhi, for bibliography and abstracts.

Inmed plan will publish common standards of plant nomenclature, establish a network graphic centre for members, liaison with international database and provide periodic newsletters, technical support and training on database management.

FRLHT's revitalisation programme aims to make village communities more self-reliant in health care in rural India, and help traditional Indian medical systems make original contribution in the fields of skin, liver and gastrointestinal disorders, ophthopaedics, opthalmology and mental health. The tasks on its agenda are to document Indian health care traditions, conserve flora on which indigenous medicines are based and set up cottage scale entrepreneurial herbal medicine production units in rural areas.

Considerable progress in health care has been made; however, in the context of the vastness of the country it is much below what is required. It is a sad comment that over the years investment in health, though rising in amount has actually declined as a percentage of total plan investment from 3.30 in the First Plan to 1.88 in the Seventh. The family planning efforts have not resulted in the required decline in populawtion. The state of health of the general population and services offered is not adequate. The bed population ratio in the country is 0.75 per 1000 population, as on January 1, 1993. Health for All by 2000 AD seems an elusive goal.

6

SPACE RESEARCH

The Indian space programme which began in a modest way in 1963 is now set to take on complex missions like building Polar Satellite Launch Vehicle (PSLV), Indian Remote Sensing Satellites (IRS) and indigenous INSAT satellites.

Within ten years after the launch of the first Indian scientific satellite Aryabhata, designed and fabricated by engineers of the Indian Space Research Organisation (ISRO) from a Soviet Cosmodrome in 1975, Indian space scientists have built two earth observation satellites Bhaskara I and II, and an experimental communication satellite APPLE, besides four small Rohini satellites for the indigenous SLV-3 launch vehicle. The principal objectives of the space programme are to develop indigenous competence in designing and building sophisticated hardware involved in space technology including rockets and satellites for scientific research and practical applications.

The space scientists are also engaged in creating the country's own launch facilities by building satellite launch vehicles. Four SLV-3s (Satellite Launch Vehicle) have been tested for putting the Rohini satellites into orbit. The first successful launch in July 1980 put India firmly on the space map to become the sixth nation in the world to have satellite launch capability. Plans are on to build rockets to launch heavier satellites.

The primary objective of the Space Programme is to set up self-reliant operation of space services. The main thrusts of the Programme are: (i) satellite-based resources survey and environmental monitoring, (ii) satellite communications, (iii) meteorological applications, and (iv) development of indigenous satellite launch vehicles and associated ground segment for providing these space based services. The S&T projects of the Department of Space (DoS) fall under the following categories: (i) National Mission/Mission-mode projects; (ii) projects of DoS; and (iii) DoS-sponsored projects in universities and other institutions.

The Space Commission which was set up in 1972 is the nodal agency for coordinating R&D activities in space science and technology. The executive wing of the Commission is the Department of Space (DoS) which operates through a major setup called Indian Space Research Organization (ISRO), Bangalore and four other independent bodies, viz. Indian National Satellite Space Segment Project (INSAT-SSP), National Natural Resource Management System (NNRMS), National Remote Sensing Agency (NRSA) and the Physical Research Laboratory (PRL), Ahmedabad besides sponsoring research projects in other institutions.

Research and development activities under the space programme are carried out in various centres/units of ISRO/DoS. The major projects of ISRO are executed by identified 'lead centres' depending upon the technology base established in them.

Vikram Sarabhai Space Centre (VSSC) is located at Thumba in Trivandrum. It has establishments at valiamala for major facilities of PSLV project and at Vottiyoorkavu for development of composites. An ammonium perchlorate experiment plant is located at Alwaye, VSSC is the largest ISRO Centre. It provides technology base for the country's indigenous satellite launch vehicle development efforts. The Centre is also responsible for the development of several spacecraft subsystems. VSSC is the lead centre for major projects such as ASLV, PSLV and GSLV. Supporting these projects are specialised R and D groups in avionics and mission dynamics, solid propulsion, propellants and chemicals, materials and mechanical systems, etc. VSSC also supports Rohini Sounding Rocket (RSR) Programme. The Space Physics Laboratory (SPL) is also located in VSSC.

ISRO Satellite Centre (ISAC) at Bangalore is the lead centre for satellite projects of ISRO. The Centre is responsible for design, fabrication, testing and management of satellite systems for scientific, technological and application missions. Aryabhata, Bhaskara,

APPLE and IRS-IA satellites were built here. At present IRS and INSAT series of satellites are being built in this centre.

SHAR Centre is located in Sriharikota Island on the east coast of Andhra Pradesh. It is the main launch base for launching satellite launch vehicles and sounding rockets. The country's largest, solid propellant space booster plant (SPROB) are located here. The Centre also has static test and evaluation complex (STEX) and computerised data processing facilities to support indigenous launch vehicle missions.

The Liquid Propulsion System Centre has facilities for development of spacecraft propulsion system at Bangalore and for launch vehicle liquid propulsion system at Valiamala near Trivandrum. The major test facilities are located at Mahendragiri (Tamil Nadu). The Centre also manufactures Precision Transducers at Bangalore.

Space Applications Centre at Ahmedabad is involved in research and development in space applications. It has the primary responsibility to conceptualise, plan and execute projects and research programmes leading to practical use of space technology. The activities include satellite based telecommunications and TV and remote sensing for natural resources survey and management, environmental monitoring, meteorology and geodesy. The Centre is organised functionally into satellite communications, remote sensing and microwave remote sensing. Support services are provided by technical services group consisting of environmental test facilities, electronics and mechanical fabrication facilities and reliability and quality assurance group.

Developmental and Educational Communication Unit (DECU) at Ahmedabad is involved in various areas of software research such as television programme production, policy studies and research in society technology interaction. Major tasks of DECU include: (a) Kheda Communication Project, (b) training of staff especially for Doordarshan, (c) policy and technoeconomic studies related to applications of space technology.

ISRO Telemetry Tracking and Command Network (ISTRAC), with its headquarters at Bangalore, operates a network of ground stations to provide tracking, telemetry and command (TTC) support for launch vehicle and satellite missions of ISRO. Development of software packages and operation of network station for present and future missions are the responsibilities of ISTRAC. It has a Multi-Mission Spacecraft Control Centre at Bangalore. The unit has established an S.band receive only station in Mauritius to support the IRS mission. ISTRAC also operates Satellite Tracking and Ranging Station (STARS) at Kavalur in Tamil Nadu.

ISRO Inertial System Unit which is located at Trivandrum is responsible for design and development and inertial systems both for satellites and launch vehicles such as momentum/reaction wheels, solar array drives, rate gyro packages, etc.

National Remote Sensing Agency (NRSA) at Hyderabad is an autonomous institution supported by the DoS. The agency has facilities for surveying, identifying, classifying and monitoring earth resources using aerial and satellite data. The analysis, computation and interpretation facilities are located in Hyderabad and satellite reception station is located at Shadnagar near Hyderabad. The Earth Station has facilities to receive data from remote sensing satellites such as LANDSAT, NOAA, SPOT and IRS, NRSA also operates Indian Institute of Remote Sensing (IIRS) at Dehradun.

Physical Research Laboratory (PRL) at Ahmedabad is the premier national centre for research in space and allied science. It is an autonomous institution supported mainly by the DoS. PRL is also entrusted with management of Udaipur solar observatory. Its main research programmes are in solar planetary physics, infrared astronomy, geo-cosmophysics, theoretical physics including climatology, plasma physics and laboratory astrophysics, archaeology and hydrology. Its 1.2 m reflecting IT telescope installed at Gurushikhar near Mount Abu is the largest of its kind in the country.

The Indian space programme took a gland leap forward during the eighties elevating India to an important member of the international space club. Many prestigious programmes like the Indian National Satellite System (INSAT) and remote sensing through satellite got underway in that period. It is, however, obvious that while significant progress has been achieved in the area of building, deployment and management of communications and remote sensing satellites, the same success has not attended ISRO's attempts to develop launch vehicle technology.

The milestones in India's space programme — success and failures — are briefly surveyed below :-

1962 : Indian National Committee for Space Research (INSOSPAR) formed by the Department of Atomic Energy.

1963 : Thumba Equatorial Rocket Launch Station (TERLS) established to measure under atmospheric parameters, particularly equatorial electroject.

1965 : The Space Science and Technology Centre set up at Thumba as a research and developments laboratory in space technology for achieving self-reliance in the fields.

1967 : Rohini-75, first Indian rocket developed at SSTC, successfully launched.

1969 : Indian Space Research Organisation set up for the peaceful exploration of space.

1972 : Airborne remote sensing experiments commence to survey earth resources.

1975 : The first Indian Scientific satellite Aryabhata launched on April 19, from Soviet Union.

1975-76 : The first major space application programme, Satellite Instructional Television Experiment (SITE), carried out between August 1975 and July 1976 using the US satellite ATS.

1977 : The Satellite Telecommunications Experiment Project (STEP) conducted from mid 1977 to 1979 using the Franco-German satellite 'Symphonie".

1979 : The Second Indian satellite Bhaskara-I, a satellite for earth observation, launched on June 7, from Soviet Union.

1980 : SLV-3 satellite launch vehicle puts 35 kg Rohini satellite (for studying the parameters of the rocket itself) into near earth elliptical orbit from Sriharikota on July 18.

1981 : SLV-3 carrying a 35-kg Rohini satellite launched from Sriharikota on May 31 ends its mission in nine days due to low altitude problems.

APPLE (Ariane Passenger Pay-load Experiment) First indigenously built experimental communication satellite, launched from Kourou, French Guyana, on June 19, Aim : acquiring first-hand experience in designing, building, launching and operating a three-axis stabilised spacecraft in geostationary orbit.

Bhaskara-II, earth observation satellite launched from Soviet Union on November 20. Very useful in land cover mapping, snow mapping and geology and vegetation coverage of the country.

1982 : INSAT-1A launched from United States by US Delta rocket on April 10. Mission aborted in September due to snags.

1983 : SLV-3 D2 lifts off from Sriharikota on April 17 with 42 kg Rohini Satellite. Imageries sent by the satellite helpful in study of the Bhadravati forests in Karnataka, snow cover in Sutlej and vegetation in northern Andhra Pradesh. Successful completion of mission in September 1984.

INSAT-1B launched by US space shuttle Challenger on August 30, A success.

1984 : The first joint Indo-Soviet manned space odyssey commences on April 3. Sqn. Ldr. Rakesh Sharma becomes the first Indian cosmonaut.

1987 : ASLV-D1 lifts off from Sriharikota with a Stretched Rohini Satellite on March 24. The mission fails.

1988 : Indian remote sensing satellite IRS-1A launched from Soviet Union on March 19. A great success. Imageries of high quality and precision obtained, comparable to those sent from sophisticated satellites like LANDSAT.

ASLV-D2 launched from Sriharikota on July 13. Fails

INSAT-1C launched from French Guyana on July 22. Loses earth lock on January 1, 1989.

1989 : INSAT-1D suffers damage while mating with its launcher on June 20. Launch postponed.

1990 : INSAT-1D launched on June 12 by the Delta 4925 launch vehicle from Florida, USA.

1991 : IRS-1B launched on August 29 from Baikanour in Kazakhstan.

1992 : The Augmented Satellite Launch Vehicle (ASLV-D3) launched by the Ariane rocket of the European Space Agency from Kourou, French Guyana. Successful.

1993 : INSAT-2B, second indigenous multipurpose satellite launched by the Ariane rocket from Kourou, French Guyana, on July 23. Successful.

PSLV launched on September 20 fails maiden attempt. Plunges into the sea after some minutes of smooth flight. Thus 850-kg IRS-1E satellite not placed in orbit.

1994 : Successful launch of ASLV-D4 from Sriharikota on May 4. Places the 113-kg SROSS-C2 satellite in orbit.

In October, the second developmental flight of the PSLV-D2 from Sriharikota proves successful. IRS-P2 put into orbit.

1995 : INSAT-2C, the heaviest of Indian communications satellites, launched by the Ariane rocket from Kourou in French Guyana, in December. INSAT-2C does not have meteorological payload.

The IRS-1C launched in December from the Baikonur Cosmodrome in Kazakhstan by a Russian rocket.

1996 : PSLV-D3 launched in March from Sriharikota; places IRS-P3 in orbit.

The Indian National Satellite (INSAT) system is a multi agency, multipurpose operational satellite system for domestic telecommunications, meteorological observations and data relay, nation-wide direct satellite television broadcasting and nation-wide radio and television programme distribution. It is a joint venture of DoS, Department of Telecommunications (DoT), Indian Meteorological Department (IMD), All India Radio (AIR) and Doordarshan. The Department of Space has direct responsibility for the establishment and operation of INSAT space segment.

The INSAT-1 series, custom-built to Indian specifications by the American company, Ford Aerospace, only partially served the country's needs. INSAT-1A and INSAT-1C were failures. INSAT-1B completed its designed life of seven years in August 1990. INSAT-1D has been the mainstay of the INSAT-1 series.

INSAT-2A was the first in the series of five satellites to be built in the country by ISRO to replace the ageing INSAT-1D and keep the vital communication lines open round-the-clock. It is programmed to provide an array of services — more telecom circuits, television and radio broadcasting facilities, improved weather service and an experimental satellite-aided search and rescue service. INSAT-2B enabled Doordarshan to expand its network. The extended C-band transponders are at the disposal of DoT and business subscribers for establishing data networks. For the Indian Meteorological Department, INSAT-2B mainly serves as an in-orbit back-up for collection and dissemination of meteorological data via the data relay transponder. The satellites's 406 mega hertz search and rescue package relays to a central place any emergency alert message originating from the seas. The worldwide service uses four low-earth orbit satellites which pick up distress signals from standard ships, aircrafts or groups of people help up in remote areas such as mountains and forests and transmit them to the nearest rescue centre. The rescue efforts are then set in motion locally. Using a geostationary statellite will help cut down the time spent in alerting the rescue centres. INSAT-2 will activate the disaster warning system located on the east coast and provide accurate information on approching cyclones.

An important payload on INSAT-2A and 2B satellites is the indigenously built very high resolution radiometer (VHRR) which provide pictures of the cloud cover, every three hours,over the country. These pictures, along with other information such as wind velocity in the upper atmosphere, sea surface temperature and precipitation play a key role in predicting the weather over the country and also help in developing better mathematical prediction models.

INSAT-2C has been lodged next to INSAT-2B so that for a ground station the two together appear as one big satellite. This 'co-location' of the satellites at 93.5^{0}E has become necessary because of congestion at that height with over 160 geostationary satellites in orbit. With INSAT-2C, which is expected to have a life of 10 years, operation with co-located satellites will be the first ever for the Indian space programme.

Since INSAT-2A and 2B provide enough capacity for meteorological data, the meteorological payload has been done away with in INSAT-2C and hence, unlike the earlier INSAT satellites, it is a dedicated communication satellite. The weight gained by doing away with the meteorological payload, has been used to augment the communication channel capacity. INSAT-2C is the first Indian satellite to carry the Ku-band transponders. The Ku-band has the advantage that a small roof-top terminal is enough to receive and send signals. Also because it does not carry a meteorological payload. INSAT-2C is symmetrical in the North-South direction. The solar panels stretch out symmetrically on either side. The satellite's three Ku-band transponders will be used for business communication in the four metropolitan cities and the two high-power C-band transponders for beaming television programmes to a large number of countries stretching from northeast Africa, Central Asia through the Gulf region to southeast Asia. The S-band transponder for Mobile Satellite Service (MSS) introduced for the first time, will open a new chapter for cargo and shipping industry, enabling them to communicate while on the move. The coverage of this beam will be wider in order to provide the service over India and over maritime zones around India. Thus INSAT-2C will expand the range of satellite services already available to India apart from firmly entrenching its presence in space.

The INSAT system has had tremendous impact on telecommunication, weather forecasting and television expansion. In the words of U.R. Rao, former Chairman of ISRO, "the rapid communication revolution initiated in the country through INSATs has provided a new dimensions to our national communication network. Over 5000 two-way speech circuits covering 140 routes, amounting to 1,20,000 route km of communication network, has enabled us to connect for the first time, even remote rural areas and off-shore islands with the mainstream of the nation using Low Cost Terminals (LCTs). The nation-wide geographic reach of INSAT system is being advantageously used for a variety of applications such as administrative, business and computer networking using small terminals. Dedicated specialised services, such as PTI news services, facsimile transmission, emergenecy communication, highly innovative rural telegraphy, NICNET network and Remote Area Business and Message Network (RABMN) with a large number of VSATs have created an unprecedented demand for further expansion of these services and thus a larger share of the space segment capacity. The entire radio networking of all radio broadcasts in the country is now through INSAT. Keeping in mind the emerging scenario of personalised communications and mobile communication, high power ku band transponders have been added to INSAT-2C and 2D satellites, in addition to the use of data and video compression capabilities and advanced modulation techniques. The most dramatic impact of INSAT has undoubtedly been in the rapid expansion of TV dissemination in the country which has resulted in providing access to over 85 per cent of the country's population with 540 TV transmitters as against less than 10 in the pre-INSAT era. The regional networking services, introduction of new metro channels, addition of satellite news gathering capability in real time have also given a new fillip to the electronic industry as a whole and in particular to the development of highly sophisticated ground systems and terminals." Weather forecasting has received an impetus with the collection of remote area meteorological data from unattended platforms.

Education can be spread through satellite networks. ISRO has conducted selected large-scale experiments by developing special inexpensive talk-back facilities. The interactive educational system was successfully experimented in Rupal, a remote village in Gujarat, where local teachers and leaders were able to interactively communicate with teachers at the SAC and Gujarat Vidyapith at Ahmedabad. ISRO and UGC have conducted a successful experiment enabling students in universities to interact, in real time, with experts through specially developed talk-back facility. An innovative interactive training programme for farmers in Haryana has also been conducted, as also an experiment in continuing education for industrial workers.

Eradication of illiteracy and promotion of scientific ideas indeed the very development of the country can be tackled with the help of dedicated satellites specially programmed to disseminate region-specific information to each language group / ethnic community in rural India. ISRO has come up with the idea of dedicated GRAMSAT satellites carrying six to eight high powered C-band transponders which together with video compression techniques can disseminate region and culture specific audio-visual programmes of relevance in each of the regional languages through rebroadcast mode on an ordinary TV set. The high power in the C band will enable even remote area viewers outside the reach of TV transmitters to receive programmes of their choice in a direct reception mode with a simple dish antenna.

The term remote sensing refers to the process of sensing, identifying and delineating various objects on ground from a distance without coming into direct physical contact with them. It essentially involves measuring the electromagnetic radiation of the sun that is reflected, scattered or re-emitted by the objects on the surface of the earth. It is possible to identify surface objects from airborne or space platform on the basis of the reflection, of the electromagnetic radiation. Remote sensing can, therefore, provide reliable and timely information on natural resources, both renewable and non-renewable. Given India's size and its resource diversity, generating up-to-date information on terrestrial resources and environment is an essential element in the planning process which would be enormously difficult through ground based survey mechanisms. However, ground data is also important to provide sampling verifications of remote sensed data because often different objects give out similar or overlapping spectral responses.

Space borne remote sensing through satellites constitutes a far greater improvement over the air-borne variety because of its unique ability to provide synoptic view of large areas as well as speedy acquisition and dissemination of meteorological, geological, geographical and ecological data. With a very short turnaround time, such data make important inputs to resource management and planning. For example, a few days of imagery can cover the entire forestry of the country which could otherwise take several years. Not only that, it provides information on phenomena of transient nature such as agriculture, drought and floods. Furthermore satellite-based remote sensing is much cheaper than ground-based or even airborne methods.

ISRO and the Indian Council of Agricultural Research conducted during 1974-75 a joing experiment called the Agricultural Resources Inventory and Survey Experiment (ARISE), employing remote sensing techniques to assess crops and land-use pattern in Anantapur district of Andhra Pradesh and Patiala District of Punjab.

Indian experimental satellites, Bhaskara I and II carried out remote sensing for land cover mapping, geology and vegetation cover of the country. The Indian Remote Sensing (IRS) satellites are state-of-the-art satellites, indigenously designed and put into polar orbit for useful remote sensing applications.

ISR-1A and IRS-1B provide multi-spectral imageries with a resolution of about 36 metres and a repetivity of 11 days (together), vital information on forests, environment, agricultural crops, soil conditions, waste-land identification, ocean resource development and mineral exploration, land use and monitoring of water resources. IRS-1C incorporates advanced features such as LISS-III camera, a panchromatic (PAN) camera which has a ground resolution of less than six metres, and a wide field sensor (WIFS) enabling better spectral and spatial resolutions than its predecessors, and thus enhancing the application potential of these satellites. The PAN camera collects light from most of the visible spectrum and is useful in mapping. WIFS makes it possible to revisit the same area within five days. IRS-P2 and IRS-P3 have now joined the other remote-sensing satellites to enhance the data availability. The IRS-P3 has a WIFS and a short-wave infrared (SWIR) band. The satellite is to help in assessing crop conditions and atmospheric phenomena. It will also give early warnings of floods and help to assess flood damage and study certain types of rocks. It also carries a payload of the German space agency.

Microwave remote sensing which can even see through cloud cover is a thing of subsequent generation of satellites for which the ISRO is carrying out air-borne and space-borne experiments abroad the European remote sensing satellite ERS-1, launched in 1991. In microwave remote sensing, which is different from the method of IRS-type satellites, radar beams are bounced off the land, picked up and analysed.

Work on microwave remote sensing has started with a project launched to develop an air-borne synthetic aperture radar. Microwave remote sensing satellite can be used for fish school location, coastal zones management, ocean geoid and ocean surface height determination, understanding of atmosphere boundary layer and dynamics on the ocean surface, etc.

To utilise the data and data products for applications, several Central and State government agencies have set up remote sensing centres for agricultural crop acreage and yield estimation, assessment of pest infestation in crops, drought monitoring and assessment, flood control and damage assessment, land use and land cover mapping for agroclimatic zone planning, wasteland management (several types of wastelands have been identified), water resource management (particularity, ground water surveys in more than 400 districts), urban development, mineral prospecting, forest resources survey, impact of mining, climate geosphere studies, coastal environmental evaluation, etc. Ground water surveys have in some areas, notably in Rajasthan, shown 90 per cent success. IRS also has played an important role during cyclones as well as in tracking oil slick movements.

A major application of IRS data is in the IMSD or Integrated Mission for Sustainable Development which was begun in 1992. Under IMSD, more than 174 districts have been identified so far for generation of action plans for locale specific development. Six such plans have already been taken up for implementation.

The National Natural Resources Management System (NNRMS) with its headquarters in Bangalore is a unique system in which data from remote sensing are integrated with conventional sources for effectively managing the country's natural resources. NNRMS undertakes natural resources survey and monitors projects in collaboration with Central and State Agencies.

The IRS satellites are sending plenty of data but some reservations exist about the use of this data. Utilisation of the data is still in the experimental stage. Some users complain that the data supplied by the NRSA is poorly composed. The data has, however, been of great use to the Forest Survey of India and the Union Water Resources Ministry. The Defence Ministry, concerned with security issues, screens all data received by the NRSA and masks parts of the images — this is a handicap to users looking for a complete picture. It is also pointed out that the sheer volume of data received is making it difficult to store. In the circumstances, more efforts should be made to utilise effectively all the data being beamed by the IRS satellites.

Stretched Rohini Satellite Series (SROSS) Programme is directed towards supporting missions using 150 kilogram class satellites for space science, technology and applications studies.

The indigenous capability for the development of satellite launch vehicle was demonstrated through the first successful launch of SLV-3 in July, 1980, carrying the 40-kilogram Rohini Satellite.

India became the sixth country to have the capability to orbit her own satellite. The 22.7 m-tall SLV-3 was an all solid four stage vehicle and had a lift off-weight of 17 tonnes. Two more launches of SLV-3s were conducted in May 1981 and April 1983, with the Rohini satellites on board carrying application oriented solid state imaging sensors. The efforts in launch vehicle development are now concentrated to develop enhanced capability towards realising launchers for the Indian remote sensing satellites and, subsequently, for communication satellites.

The augmented satellite launch vehicle (ASLV), basically derived from SLV-3, was originally meant for putting 150 kilogram class technological / scientific pay-loads into near circular orbit. The first two developmental flights of ASLV failed but they provided valuable inputs for the future flights and also validated a number of technologies. The third development flight of the ASLV lifted off to a perfect launch. It released a 106-kg satellite stretched Rohini series (SROSS-C) into orbit. However, the success of the mission was partial; the satellite was put in a sub-optimal orbit, because of a problem in the altitude control system of the final fourth stage of the launch vehicle (AS4). As a result, the satellite lasted for only 55 days instead of the targeted 180 days and burnt up in the atmosphere.

Then on May 4, 1994 came the second consecutive success of the ASLV when ASLV-D4, launched from Sriharikota, placed the 113 kg SROSS-C2 satellite into orbit. The satellite should have a life span of three years. It has two sets of equipment on board; a gamma ray analyser and a retarding potential analyser to study the ionosphere.

The ASLV is a five-stage rocket using solid fuel. The development of the solid fuel, particularly hydroxy-terminated polybutadiene (HTPB) and lactone-terminated polybutadiene (LTPB or HEF-20) used in ASLV, is a great achievement for ISRO. The ASLV is 23.85 metre high and has a gross lift-off mass of 1.8 tonnes.

The ASLV was needed as a stepping stone to test and prove more cheaply some of the important technologies needed for the polar satellite launch vehicle (PSLV), such as closed loop guidance system, strap down inertial navigation system, bulbous heat shield of the payload bay, etc. A launch vehicle is enormously complex for, while coping with the difficulties and vagaries of flying through the earth's atmosphere, it has to keep to a narrowly defined trajectory and attain the right velocity for the satellite it carries to be placed in the correct orbit. Unlike satellites, important components for launch vehicles — let alone technology — can rarely be imported for fear that these can be used for missiles. With the Missile Technology Control Regime, such imports have become next to impossible. Classic examples are the guidance system and solid propellants. The ISRO has had to design and build its launch vehicles virtually from scratch. The result has been that the launch vehicle programme has lagged behind the satellite development.

The first developmental flight of the polar satellite launch vehicle (PSLV) failed but the second one proved a grand success for India's space programme. This achievement of ISRO's enabled India to enter the world launch vehicle market. The PSLV, the country's first operational launch vehicle, is a four-stage rocket. The first stage is a solid propellant motor with six SLV-3 first stage motors strapped to it. The second stage is based on the liquid engine technology of the European launch vehicle, Ariane, and uses liquid propellant. These stages form the core of the next generation GSLV. The third stage is a solid propellant motor and the fourth stage, a liquid propellant stage. A closed loop guidance system with on-board processor is employed for the vehicle.

The rocket put the remote sensing satellite IRS-P2 into orbit at an altitude of 820 km. This was the first IRS satellite put into orbit by the PSLV.

In 1996 the PSLV-D3 successfully deployed the 930 kg IRS-P3 satellite into a near polar sun-synchronous orbit. The PSLV-D3 had for its first stage one of the largest solid propellant boosters in the world. Indigenously produced hydroxyl-terminated poly-butadine (HTPB) fuel powered it and ammonium perchlorate was the oxidiser. The second stage employed the indigenously produced Vikas engine and a liquid propellant — unsymmetrical dimethyl hydrazine (UDMH) — as fuel and nitrogen tetroxide as oxidiser. The third stage used HTPB-based solid propellant. In the fourth and terminal stage, a twin-engine configuration used liquid propellant monomethyl hydrazinic and another grade of nitrogen tetroxide as oxidiser.

The PSLV could be used commercially to launch a variety of polar orbiting satellites. It could launch low earth orbit (LEO) satellites for which there could be great demand with the growth of global networks for hand-held mobile telephones.

The geo-synchronous satellite vehicle (GSLV) is to be 51m tall, weighing 400 tonnes. It will place a payload of over two tonnes in geostationary transfer orbit. The GSLV will be developed in two phases. In the first, GSLV Mark-I having the core vehicle as PSLV and with four liquid second stages of PSLV replacing the solid motors as first stage strapons will be developed. In the second, Mark-II will be developed which will have a cryo-stage replacing the upper two stages of Mark-I to realise a payload capability of 2000-2500 kg geosynchronous transfer orbit.

The Cryogenic engine is crucial to the development of GSLV. Cryogenics refer to technology of sub-zero temperatures and cryogenic engines use liquid oxygen as the oxidiser and liquid hydrogen as the fuel. Use of the liquified fuel instead of solid fuels or earth-stroke liquid fuels like kerosene helps the engine produce greater thrust for the same quantity of propellant. The engine's capability is thus of placing a far heavier payload.

The American Atlas-Centaur rocket was, thirty years ago, the first to use cryogenic rocket technology. The technology is complicated; use of special insulation and materials that can withstand temperatures of minus 200 degrees or less (the boilding point of liquid hydrogen is minus 253 degrees C) is necessary. The turbines and pumps delivering the propellants to the combustion chamber in the cryogenic engine have to be designed out of special materials.

India was to acquire the cryogenic engine and technology from the Russian space agency, Glavkosmos. A hitch developed when under US pressure and the Missile Technology Control Regime (MTCR) restrictions, Russia refused to honour the original contract under which India had to give Rs. 235 crore to gain three cryogenic boosters and the entire knowhow from Russia by 1995. This forced ISRO to go ahead on its own to develop a cryogenic engine though it would have taken many years to manufacture such an engine for rocket launches. But renewed negotiations in early 1994 between ISRO and Glavkosmos resulted in a resettlement providing for four mock-up models in lieu of the technology, besides the three boosters. ISRO scientists have, however, managed to design and fabricate a small one-tonne cryogenic engine.

In 1994, India successfully launched the indigenously developed cryogenic payload for measuring depletion of the ozone layer and presence of greenhouse warming substances in the atmosphere.

The Sponsored Research Programme (RESPOND) of ISRO, initiated in 1970, has established a strong interaction between ISRO and academic institutions to carry out research, developmental and educational activities. The aim is to develop a large research base at academic institutions, for manpower training, building infrastructure facilities and nurturing indigenous capability to support the space research programme.

The Indian space programme does not work in isolation. Many space-oriented technologies have been transferred to Indian industries for non-space use. The partnership between ISRO and Indian Industry involves transfer of advanced technologies developed by ISRO to industry and provision of technological consultancy from space programme to industry on the one hand and utilisation of industry's own technological potential and expertise by the space programme on the other. In this context it may be pointed out that Indian industry had a major role in the PSLV project. The special maraging steel was manufactured by MIDHANI; the motor cases by Walchand and Larsen and Toubro; the liquid Vikas engine by Godrej and Boyce, and MTAR; liquid propellants by Andhra Sugar and Hindustan Organic; solid propellants by NOCIL; the 12-storey tall moving tower by Triveni Structurals.

ISRO's Technology Transfer Scheme promotes and supports commercialisation of technologies developed by ISRO for applications in various fields. The main aims of this scheme are ; (a) meeting requirements of space programmes and projects through buy back of products produced by industry as a result of technology transfer; (b) servicing rapidly expanding space applications markets in India generated by space programme in satellite communications, television and radio broadcasting, meteorological services, remote sensing for natural resources survey and management; and (c) exploiting full potential of indigenous technologies developed by ISRO for multifarious spin-off applications. The role of Indian industry in participating in the projects has shown continuing growth both in terms of technological complexity and quantum of work. Certain premier industries in the field of aeronautics communications and engineering have now set up their own specialised departments to deal with space products. These include rocket hardware, spacecraft and launch vehicle structures, propulsion engine components, propellants, etc.

Examples of technologies transferred are fibre glass technology, pressure transducers modified to be used for monitoring blood pressure during heart surgery, chemicals like pylol, computer software, high silica cloth, olfex powder for extinguishing oil fires. The major areas benefiting from the spin-offs from space research are polymers, chemicals and special materials, telecommunications, pyroignited systems, optics, opto-electronics and opto-mechanical equipment. Industries have been able to perfect innovative materials like high precision alloys, making use of ISRO's technologies with the help of ISRO's know-how and consultancy. Tata Telecom Ltd. was able to produce print circuit boards of international specifications, and a digital coding and decoding system with increased data carrying capacity.

Antrix Corporation Ltd., a wholly government-owned company, has been formed in Bangalore to market technologies and services relating to assembly and sub-assemblies of satellite systems. It has a contract from IMMARSAT to study LEO satellite systems for the INMARSAT-P programme and also develop antenna for handheld INMARSAT phones.

Antrix-EOSAT Agreement A 10-year contract was signed by the Antrix Corporation of India and the American satellite data company, EOSAT in 1995 to enable commercial distribution of data collected by Indian remote sensing satellites. EOSAT will distribute the data globally.

ISRO-Intelsat Agreement under a 10-year agreement signed on January 30, 1995, the Indian Space Research Organisation will lease out to Intelsat a part of the capacity of INSAT-2E — to be launched in 1997.

7

ADVANCED MATERIALS

India now runs neck and neck with technologically advanced countries in the development of high temperature superconductors with zero electrical resistance that can revolutionise the electronics and computer industry. The country can now produce metallic glass, a new class of material poised for revolutionising electrical industry and has made a major breakthrough in the design and development of optical fibres which have enormous promise in the field of communication, medicine and industry.

A significant and major achievement in superconductor research has been made by scientists at the Indian Institute of Technology, Madras, National Physical Laboratory, New Delhi, Bhabha Atomic Research Centre, Bombay ; Tata Institute of Fundamental Research, Bombay and the Indira Gandhi Centre for Atomic Research (IGCAR), Kalpakkam.

While in Japan, the United States and China, scientists have developed a new class of materials exhibiting superconductivity at 334 degree K (Kelvin is a unit of absolute temperature the coldest temperature conceivable), Indian scientists are very close in the race and have achieved the feat at a close of 300 degree K (27 degree celsius).

The Madras Group of Scientists working in the IIT and the IGCAR have developed a variety of new materials, as many as 50 based on the rare earth-barium, copper, oxygen system and proved to be superconductors even at 95 degree Kelvin. Samples of National Physical Laboratory in New Delhi show superconductivity at 300 degree Kelvin. Many of the elements in the new alloys (dysporosium, holmium, e-bium, thullium and ytterbium) are abundantly available in India.

The discovery, most scientists believe, can lead to incredible savings in energy. Trains that speed hundreds of kilometres per hour on a cushion of magnetism, practical electric cars, powerful, yet smaller computers and, particle accelerators, safe reactors operating on nuclear fusion rather than fission and a host of other rewards still undreamed of will be available.

Besides, it would be possible to make wires and these materials to reduce transmission losses in electricity transfer and also to produce huge magnets for industrial applications. Again, the superconducting materials could be employed in the manufacture of supercomputers.

Over 22 research institutions in the country are working on various aspects of materials and systems for optical communication such as drawing of low-loss optical fibres, splicing and connecting techniques, modulation and demodulation techniques, multiplexing and demultiplexing techniques and system studies.

The amount of information which can be transmitted by metal conductors is related to the wavelength of the electronic signal. Optical fibre can carry up to 10,000 times more information than a copper wire of the same diameter —allowing much smaller transmission channels, or alternatively, much higher transmission rates.

Scientists of the Central Glass and Ceramic Research Institute in Calcutta have developed optical communication fibres which are made from ultrapure fused silicon tube having graded refractive index.

The Telecommunication Research Centre in Delhi has developed a terminal equipment for a 120-telephone channel system while the Indian Institute of Technology in New Delhi has a facility for testing fibres and engaged in perfecting already existing measurement techniques and trying to develop new methods to characterise fibres (pulse dispersion, attenuation refractive index profile measurement).

The Central Electronics Engineering Research Institute in Pilani has developed silicon avalanche diodes for detecting and decoding laser pulse at the receiving end while the Instruments Research and Development Establishment in Dehradun and the Central Scientific Instruments Organisation (CSIO) in Chandigarh have developed a variety of fibre optics components for various applications such as light guides of different lengths and apperture, aligned flexible fibre bundles used in signoidoscope, fibrescope and traffic signs ; solid fused bundles or face plates.

The Defence Solid State Physics Laboratory in Delhi has succeeded in fabricating the gallium arsenide lasers while the CSIO is developing a machine for drawing plastic coated silica fibre. A more sophisticated facility at the Central Glass and Ceramic Research Institute in Calcutta has already produced experimental quantities of 'graded-index' multimode fibres.

Research and development and measurement techniques related to characterisation and evaluation of communication grade fibre are being done at Indian Institute of Technology (IIT), Delhi, and Central Scientific Instrument Organisation (CSIO), Chandigarh. These institutes are capable of performing well-identified and complementary tasks. Other IITs are also in the process of setting up similar facilities.

Research in the field of launching optical signals and detection of signals is going on in several institutions of the country. Work is also going on terminals circuitries for converting source detectors into different systems-required-transmitters / receivers and also repeater stations.

IIT, Kharagpur, has recently started an M. Tech. course covering technology of optical fibre communication system with its associated devices, materials and components to meet the manpower requirement. Courses on optical communication are also being given at some other universities.

In India, the first optical fibre communication system was set up in Pune connecting Shivaji Nagar and Cantonment Exchanges. Buried two metres underground, the glass fibre of 8 Mb / s capacity imported from Japan provided 120 telephone channels and has been functioning without any fault since 1979. A prototype of 40 Mb / s capacity, the second optical fibre communication system developed by the Telecommunication Research Centre and the Indian Telephone Industry (ITI) will be commissioned by the end of this year.

With installation of this system, the country would pass through demonstration stage and enter into commercial exploitation of this technology.

Research in metallurgy in India is being carried out at Banaras Hindu University, Indian Institute of Science in Bangalore, the Defence Metallurgical Laboratory (DMRL) in Hyderabad, and Bhabha Atomic Research Centre (BARC) in Bombay. Special alloys and materials needed for aircraft and defence equipment are produced in a specially established factory in Hyderabad, known as MIDHANI.

The first ribbons of metallic glass rolled out of laboratory scale spin melt unit at BHU in 1978. Since then a few other laboratories started work on metallic glass. Two years ago eight research groups of different projects were funded by the Department of Science and Technology with the objective of developing a commercially viable technology for production types of metallic glass for applications in transformers cores. The Bharat Heavy Electricals Limited, India's largest manufacturer of electrical machineries, is involved in the research and development and hopes to commercially exploit this technology in the near future.

The nosecone of rockets made by the Indian space scientists use a technology for reinforced plastics with glass fibre. Fifteen research groups from universities and national laboratories under a project funded by DST have been carrying out for over nine years research and development of newer fibres and composites under this project. Technologies have been developed for producing glass carbon and organic fibres.

The notable achievements include development of E-glass and C-glass compositions; ceramic bushings to replace expensive platinum-rhodium for drawing glass fibres; process for producing aramid fibre similar to KEVLAR fibre, process for producing continuous carbon fibre production from special grade PAN fibre and a variety of processing techniques such as filament winding, pultrusion and products like impeller vanes, silos and slotted screen pipes.

The aluminium alloy conductor NML-PM 215, developed at the National Metallurgical Laboratory, Jamshedpur, has helped sizeable savings of copper. This alloy in the form of wire rod is being produced commercially at Bharat Aluminium Company in Korba. The Indian Cable Company in Jamshedpur is producing from the

NML alloy wires meeting the specifications of the Indian Railways. A multistrand overhead conductors made from this alloy is being evaluated by the Ministry of Defence. At the Central Glass and Ceramic Research Institute in Calcutta, a facility has recently been set up for producing in pilot scale glass wires for optical communications.

Indian metallurgists have made significant contributions towards development of materials for the defence forces. Notable among them is the technology for making composite armour for battle tanks. The armour made of ceramic sandwiched steel composite armoud called Kunchan armour is adjudged to be one of the best in the world and is to be used in the main battle tank being indigenously developed by the Defence Research and Development Organization. DMRL has also developed light weight ceramic armour against small arms fire and monoliothic steel armour for infantry combat vehicles. A major achievement of defence metallurgists is the development of armour piercing anti-tank projectile, an alloy material for use as penetrators. A Rs. 700 million factory to produce these projectiles is being set up in Tiruchirapalli in south India.

Scientists are also undertaking research on brake pads for various types in Air Force and civilian aircraft. DMRL has already developed the brake pad assembly of MIG-21 aircraft wheels comprising a bimetallic stator and a friction material rotor. The development was necessitated by an acute shortage of the Russian component in the Air Force inventories. The bi-metallic stator and the metallic-ceramic friction rotor are being produced at the Hindustan Aeronautics Ltd. (HAL), Koraput, and HAL, Hyderabad, respectively, HAL, Hyderabad, is also planning for a production of 10 lakhs break pad rotor assemblies per annum in the near future at a cost of Rs. 20 crores.

DMRL, the pride of Indian metallurgists, is in fact one of the best equipped laboratories in the world. The structural metallurgy laboratory has optical microscopes, X-ray diffractometers, an X-ray fluoroscene spectrometer and a most advanced electron microprobe analyses with computer output. Among the latest acquisitions is the scanning Auger electron microprobe. The mechanical metallurgy laboratory is also comprehensively equipped; Instron, a hot torsin testing machine and a whole range of devices for fatigue and creep testing. A closed-loop servchydraulic universal testing machine has just been acquired. The non-destructive testing laboratory has ultrasonic flaw detectors, magnetic track detectors and instruments for gamma-ray and X-ray radiography.

Analysis of chemical composition of even microscopic amounts of sample materials can be made to a few parts per billion by instruments that include atomic absorption spectrometers inductively coupled plasma emission spectrometer, automatic recording spectroscope, Echelle D.C. and plasma spectrometer, a computer controlled oxygen and nitrogen analyser and analysers equipped with digital display for detecting carbon, sulphur and hydrogen. The power metallurgy laboratory is completed with a full range of presses and furnaces and full facilities for making dies. It can boast of a large-size organ atomiser, cold isostatic and hot isostatic presses suitable for development of production scale processes in powder technology. The hot isostatic press, which would be the pride of any powder metallurgy research laboratory anywhere, is the only such press in the country. It can produce directly from powder components almost in finished form, bond dissimilar materials, heat defects in metal casting or give a fresh lease to used up components like turbine blades that before had to be replaced and discarded. Among the other proud possessions of the gas turbine materials group are a 'burner rig' that can simulate temperature condition in a gas turbine and a plasma coating unit for protective coating of components against hot corrosion. Soon to be installed are two multi-kilowatt laser units for the most up-to-date development in processing metals.

There is a vigorously active group working on ceramic materials for applications in electronics and high temperature structures. DMRL has an automatic press, an extrusion press and high temperature furnaces to synthesise ceramic components for special applications.

Facilities exist for melting magnetic materials in bulk quantities, pressing and heat treatment of these materials in powerful magnetic fields and for evaluating the magnetic properties of the final products. There are also facilities for investment casting of turbine blades and other precision components. DMRL can now by directional solidification make fully columnar cast superalloy and even single crystal blades for turbo-jet engines. DMRL is now on the way to successfully developing the technology for investment-cast, directionally solidified, internally cooled turbine blades for a new aircraft planned for defence production in India.

DMRL is also working on the development of high temperature and high strength titanium alloys that could reduce the weight of jet engines, making them higher and more efficient and is all set to produce titanium metal in large scale batches using indigenous know-how in its demonstration plant.

Casting of titanium was once considered difficult if not impossible because of the molten metal's aggressive nature. DMRL scientists are trying to break this barrier by directly casting titanium alloys into required shapes. The latest forming techniques which are cost effective and energy saving are superplastic forming and diffusion bonding. The laboratory has identified titanium alloy that can be formed superplastically and have designed and built a special diffusion bonding set up.

The laboratory has developed the technology for liquid forging of various aluminium alloys, copper alloys and ferrous alloys.

The importance of special materials with high levels of purity in the successful implementation of our nuclear power programme was realised during the early days of the planning of our nuclear programme. Considering the long period of research and development in the laboratories followed by setting up of prototype plants that is needed before large-scale production units can be set up, work on many of the materials needed specially for nuclear programme was initiated in the early 1950s. This foresight has gone a long way in achieving self-sufficiency in the production of important special materials needed for the nuclear programme of the country.

Metallurgy of important special materials like tantalum, niobium, beryllium rare earths, aluminium oxide, magnesium oxide ziroconium oxide, rapidly solidified metals and oxides of uranium and thorium which are of importance in the nuclear programme and in the development of space, electronics and magneto hydro-dynamics technology has been extensively studied at Bhabha Atomic Research Centre (BARC), Trombay. These studies are backed with strong support to basic research in frontline areas like phase transformations, structure property correlations, diffusion, radiation, induced changes, fracture mechanics, failure analysis, stress, corrosion plating process and sintering mechanisms.

Excellent facilities exist in BARC for carrying out advanced research and development in many branches of metallurgy like extraction physical corrosion, electrometallurgy and ceramics. The programmes at BARC have established a sound base for production, processing, fabrication and application of many special and rare materials. This work has also served as the precursor of Nuclear Fuel Complex at Hyderabad. The Beryllium Plant at Vashi, New Bombay, and the Tin Demonstration Plant were set up near Raipur.

The Beryllium Plant at New Bombay has facilities for the production of beryllium blocks and for machining of beryllium into desired components. It also produces copper beryllium alloys. These materials and products are used in the nuclear and space programmes.

The Tin Demonstration Plant, built in collaboration with the Madhya Pradesh Government at the Urla Industrial Estate near Raipur, produces about 2 tons of tin metal per year from the cassiterite mineral deposit found in that area. The successful demonstration of the basic process developed at Trombay in the pilot plant will lead to setting up larger plants to meet all the tin requirements indigenously.

An aluminium base sacrificial anode (SUPERAL) for cathodic protection of ship's hull, developed at the National Metallurgical Laboratory (NML), Jamshedpur, is being used by industry. Stable integral colour anodising and electrolytic colouring of aluminium and its alloys have also been developed by NML. Aluminium alloy filler wire for welding electrodes has been developed. The know-how for reactive filter medium for grain refining of aluminium, developed by NML has been used by the industry.

"Pyroloy 1000", a heat resistant iron alloy which can replace other costly heat resistant materials in high temperature applications, has been developed at NML. Another cast iron alloy, which replaces imported materials in abrasion wear applications such as impeller blades in shot blasting machines, chute liners for sand slingers in foundries, etc. has been developed and christened as "NML-wearnot".

Nickle-free creep resistant austenitic steel developed at NML has been used to make automotive exhaust valves. Technology for production of different type of clad-metals like copper aluminium stainless steel, mild steel, copper-mild steel, etc. has been developed at NML. Duplex, milk steel / tool bonded cutting tools leading to reduction in tool steel usage have been made by NML. Pilot level hydrostatic extruder has been used to extrude copper clad aluminium and complex profiles in copper and steel. Process for extraction of vanadium from wastes of alumina plants and from vanadifferous iron ores developed by NML has been utilised by industry. A simplified and energy saving method for recovery of chromium from chromite ores has been developed by RRL, Bhubaneshwar. The laboratory has developed another process for depositing chromium in the form of brittle sheets from chromic acid containing sulphate ions.

A pilot plant for sintering manganese ore fines has been set up by RRL, Bhubaneshwar. Under a collaborative project with industry the laboratory has assisted in setting up a pilot plant for heat hardened agglomerates from sintered iron ore fines. Technology for production of electrólytic magnesium metals from different raw materials like magnesite, carnallite, dolomite and sea bitterns has been developed by NML.

The process has been developed for manufacturing several intricate components. The technology is being transferred to Bharat Earth Movers Limited for manufacturing chosen components used in the early moving equipment which are presently being imported. This will be the first such instance of a major manufacturing facility being set up for liquid forged components which will certainly generate awareness of this manufacturing technology in the country.

In the field of fibres and composites, Indian scientist have developed newer fibres and composites having applications in industries such as aerospace, automobile, marine, chemicals, electrical and electronics. They have developed know-how for fabrication of aramid fibre, carbon fibres, optical fibres, and developed a number of resins and fibre reinforced plastic composite which may help in replacing metals because of their non-magnetic, corrosion resistant and energy efficient properties. These composites have high specific strength and specific modules and offer fibre-orientation as a new degree of freedom for the designer.

Work in India has been done in three phases — Phase I (1975-80), Phase II (1980-85) and Phase III (1985-90). In the first phase, development of fibres, which contribute mainly to the strength of a composite was taken up. The fibres selected for development were E glass fibre, polyacrylonitrile fibre, carbon fibre and aramid fibre. In addition, development of ceramic bushing for fiberising glass to replace the expensive platinum alloy bushing, fabrication of equipment like pultruder and user items like silos, was taken up. This resulted into development of E glass fibre, known as electrical grade fibre. A composition based on totally indigenous raw materials was developed by the Central Glass and Ceramic Research Institute (CGCRI) in Calcutta. The E glass produced conformed to international standards. Some inplant trials for productionisation were also conducted in cooperation with the industry.

A special grade of polyacrylonitrile fibre is the precursor for carbon fibre required for reinforcement. Copolymers and their fibres were prepared and tested in the laboratory and other studies were carried out by the Central Leather Research Institute (CLRI), Madras.

In the field of carbon fibres, the National Physical Laboratory (NPL) in New Delhi has carried out detailed laboratory studies on stabilisation by oxidation and two state carbonisation of imported special grade acrylic fibre. Special parameters of the precursor fibre, which directly influence the formation of good quality carbon fibre were studied in the laboratory. Recently the laboratory has started semi-commercial production of these fibres.

The established industrial practice of fiberising glass is through platinum alloy bushings. Laboratory studies on the development of an alumina bushing for fiberising glass at IIT, Kanpur, were successfully conducted. A glass fibre manufacturing industry is in the process of trying out the developed process as per industrial practices.

A number of products based on fibre reinforced plastics (FRP) products like silos, designed and fabricated by Vikram Sarabhai Space Centre (VSSC) which were tried by the Foodgrain Storage Authority (Warehousing Corporation) at Hapur and impeller vanes, also designed and fabricated by VSSC, were found acceptable.

In the second phase — during 1980-85 — efforts of Indian laboratories were concentrated on development of matrix resin systems and related ancillary chemicals. The methodology and philosophy of work of Phase I was continued. This phase was extended beyond 1985 to allow completion of residual work on some systems and scale-up activity of the systems.

During 1980-85, reactive polyamide curing agent for epoxy resins which imparts flexibility and consequent high impact strength to the cured resin were developed by VSSC on a 30 kg batch level. The product was evaluated by the National Aeronautical Laboratory (NAL). The process details, along with specifications, were intimated to NRDC for transfer of technology to industry.

Polyamide (condensation type) resin based on pyromellitic dianhydride is recommended for high temperature resistant electrical insulation coating/film. The resin system, developed by VSSC, was tested by an industry as an insulation coating of wire. The process details along with specification were intimated to NRDC for transfer of technology to industry.

Furfurdehyde is indigenously manufactured from renewable source materials. The alkali resistance and dielectric strength of phenolo-furfuradehyde are superior to those of general purpose phenolformaldehyde resin. One such resin system was developed by Sardar Patel University (SPU), Vallabh Vidyanagar, and large scale preparation up to 10 kg batch level was studied by VSSC. The resin was evaluated by NAL as matrix resin for glass fibre composites and its superior alkali resistance checked with carbon fibre composites. Its superiority over phenolic resin as performing resin for carbon-carbon composites of producing composites of higher density and flexural strength was established by NPL. Its superiority for making into standard grades of moulding powder and mouldability into standard mechanical and electrical goods was established by industry. The resin system has been put up for transfer to industry by the Department of Science and Technology.

Pitch impregnating resin conforming to international standards was developed by NPL and evaluated by industry. NAL evaluated carbon-carbon composites made by NPL using this pitch. The process is proposed for further study on a larger scale in the first stage by NPL and finally by RRL, Hyderabad.

Vinyl ester resin was developed by Indian Institute of Technology (IIT), Delhi, on laboratory scale initially for facilitating low temperature polymerisation of bismaleimide. Its development as matrix resin for glass fibre composites was followed up after testing by NAL and then scaled up by VSSR on a 30 kg batch level. It has been evaluated by industry for suitability, for fabrication of containers, for chemical industry.

Two grades of Bisitaconimide resins which are high temperature resistant matrix resins, high performance grade and low cost grade have been developed by VSSC and evaluated by NAL. Large scale preparation of the high performance grade was also studied by VSSC and the process parameters were finalised. Scale up of the low cost grade is in progress. A high performance grade of Bismalemide resins with high char yield value was developed by IIT, Delhi, and tested by NAL. Scale up of this system is in progress.

Enhancement of mechanical and thermal properties by reinforcing polypropylene leads to higher performance and value added products. Polypropylene reinforced with either mica fines (a waste by product in the country) or tale or calcium carbonate were developed by Indian Petro-Chemicals Ltd. (IPCL), Vadodara. These reinforced materials were found to have good processing characteristics and process reproducibility. The reinforced compounds were evaluated by NAL, Bangalore, and IIT, Bombay. Mouldability of the filled systems for products like moulded luggage, two-wheeler engine, housing for fan regulator, acid battery covers and fan blade assembly was checked and the standard of mouldability was found to be good. Fan abide assemblies moulded from reinforced polypropylene are under evaluation for performance by the industry.

Boron trifluoride-monoethylamine complex curing agent which is a special purpose delayed action type agent for epoxy resins is capable of producing high crosslink density and high order of electrical properties. It was developed by VSSC and evaluated by NAL.

HET acid based polyster resin which is a transparent flamc retardant grade polyester matrix resin for glass fibre composites was developed by Sri Ram Research Institute, Delhi, and evaluated by NAL. Reproducibility was checked on a laboratory scale and specification was drawn up. The know-how is available to entrepreneurs for scale up and transfer.

Y-methacry loxypropyltrimethoxy silance which is an effective coupling agent used extensively in good quality glass fibre reinforced polyester composites was developed by IIT, Delhi. The know-how is available for scale-up and transfer.

The current trend is not to wait for development of an asoteric new polymer to meet specific property and process requirements but to try to acquire them by blending/alloying polymers for desired property/process advantages. One such alloy of commercial polymers based on polyethylene terephthalate (PET) and polymethyl methacrylate (PMMA) was developed by National Chemical Laboratory (NCL), Pune, to give the process advantage of using of mould at low temperature. Melt-mixing of the constituents polymers was done by NCL with the twin screw extruder. The alloy was injection moulded with facilities available in industry and was tested. Dialogue with industry is on to identify likely commercial products for development from the developed alloy and their subsequent evaluation.

Cashew nut shell liquid (CNSL) is a large, low cost, indigenous vegetable source of phenolic compound and can be exploited to make many value added products. One such product, panels for housing for rural sector is an attractive application. The resin formulation based on CNSL, developed by NCL, Pune, was found to be a suitable matrix resin for bamboo mat and some indigenous low cost particulate filters. A design of a house made up of such panels for the rural sector is being worked out in consultation with architects and structural engineers to scale up the production of the resin and fabrication of the panels.

Aramid fibre (Kevlar type) of high strength, high modulus and low weight is used to produce composites that are known to fail in the ductile mode only. It finds large scale application as a superior reinforcing organic fibre in aerospace and aircraft components, personnel armour, V-belts and for reinforcement of rubber tyres of four-wheelers. Intermediate batch level study of the process of preparation of the polymer and spinning fibre from it has been in progress at NAL with equipment designed, fabricated, installed and commissioned by NAL. The process parameters are under optimisation after which the process will be ready for scale up by industry.

A pultrusion machine for fabrication of glass fibre reinforced polyester composite hollow tubes and sections was designed by CSSC. Some components were made in-house while others have been ordered from industry.

Ten products based on indigenously available glass fibres and polyester/epoxy resins were identified, designed, fabricated and tested by Poona University. All of them including paraboloid dish for solar concentrator and gasholder for gobar gas plant, are undergoing field trial.

NAL studied crack propagation by acoustic emission monitoring of an IOSIPESCU type of composite test specimen subjected to shear stress.

Regional test centres for testing resins, fibre and composites to facilitate evaluation of raw materials, composites and products by the FRP small-scale industry have been set up at IIT, Bombay, and IIT, Madras. The centres, which were founded in 1982, have now established all facilities for physical and mechanical tests. Both the centres have been testing raw materials and composites for the industry. Brochures and compendia of tests prepared by both centres for the information of industry are available. Get togethers, seminars and workshops for industry are being arranged annually by both centres to bring an awareness of the test facilities and the importance of various tests in ensuring quality control. To encourage introduction of regular testing of products in industry, particularly at the small-scale level, testing was done at both the centres free of cost for a limited period. A scheme is now being drawn up to make both the centres self-supporting and to continue to give assistance in quality control to the FRP industry.

Carbon-carbon composites (C-C composites) have shown considerable promise for use as bioimplant materials and are being investigated extensively. A plan for systematic in-vitro and in-vivo evaluation of C-C composites was drawn up by the Sree Chit, a Tirunal Institute for Medical Science and Technology, Trivandrum (SCTIMST). Some C-C composites required for this study have been made and supplied by VSSC and NPL with funds sanctioned by DST. Studies carried out by SCTIMST have shown the material to be of acceptable quality in in-vitro tests and results so far obtained in in-vivo tests point to the trend in the same directional Design of test equipment for long-term mechanical tests has been completed and fabrication is in progress.

Studies to use of composite materials in rehabilitation aids kits being jointly carried out at SDM Hospital, Jaipur, and IIT, Bombay. The first product chosen for development was a device based on floor reaction orthosis to replace metal clipers for polio-affected patients. Designed and fabricated by IIT, Bombay, and evaluated and field tested by SDM Hospital, Jaipur, the FRP aid made of carbon fibre cloth and epoxy resin has proved to be an instant success. This new device weighs about one-fourth the weight of steel calipers and is quality rigid. It is also comfortable to wear and does not required special shoes. There has been a total enthusiastic and wide acceptance by the patients (so far about seventy). A few minor improvements to increase the wear resistance of the foot portion of the device are being studied. Rehabilitation aids for above-the-knee and below-the-knee amputees are also under development with similar composite materials.

Desirability of development of such resin systems with high impact strength arose out of the knowledge gained from the study tour of project investigators during Phase II. Such resin systems find extensive applications in plotting, encapsulation and similar fields. As VSSC had already some expertise in the field related to toughening materials, a project for the development of the identified epoxy resin systems was initiated and funded at VSSC.

Development of polymers and FRP composites for specific products applications is being taken up now. It has been realised that if the development of these materials is to be meaningful, it has to be related to specific products. This orientation can be achieved only if the demands of several sectors of industry are scanned and reviewed. Further the development of a composite based product for an industry cannot lead to realistic achievement unless a total involvement of the concerned industry units is brought about during the planning of the project. Their commitment includes providing the needed direction to the R&D to meet the desired goal and giving direct or indirect financial support wherever needed. Involvement of industry right from the inception of a project is expected to ensure:

- meaningful R&D activity leading to products of direct and immediate interest to the user industry.
- participation and contribution of user and manufacturing/processing industries right from the project initiation stage to ensure correct direction of projects activity and timely inputs for the evaluation of the decided stages of development.
- preparation of an efficient and complete technology package for smooth transfer of know-how to industry and
- reduction of time lag in utilising the results of the R&D activity.

8

DEFENCE

India has had to develop special skills in national defence and aeronautics in view of growing international tension and the induction of sophisticated weaponry into neighbouring countries.

As many as forty-two laboratories of the Defence Research and Development Organisation (DRDO) have achieved significant successes in indigenisation of technology, particularly in the field of small arms, guns, ammunition, instruments, ballistics and modern explosives and developed over hundreds of sophisticated equipment to meet the growing needs of India's defence. The laboratories, in the national quest for a self-reliant defence posture, endeavour to provide the research, design and development base for meeting the needs of the armed forces — ranging from aircraft, missiles, torpedoes, radars and tanks to frozen food and nuclear medicine.

A sizeable measure of self-sufficiency has been achieved in the areas of small arms and ammunition. The Ishapore rifle, the mountain gun and the Indian field gun, mark to which DRDO made significant contributions, are now in use. A self-propelled artillary gun has also gone into production. Various types of proximity fuses and liquid fuel rocket engine have also been developed.

One of the prestigious projects of defence R&D related to the design and development of the Main Battle Tank (MBT) christened 'Arjun', the prototype of which is already developed. It has a powerful 120 mm rifled gun capable of firing 'APFSD' rounds of exceptional lethality. The tank has a most advanced gun control system and associated opto-electronics with the capability to fire on the move or a moving target. Test bed experiments, conducted at the Combat Vehicles Research and Development Establishment, Avadi, Madras, shown an overall efficiency of around 90% which is quite comparable to some of the best design developed elsewhere.

The DRDO conducts research in important disciplines like aeronautics, rockets and missiles, electronics and instrumentation, combat vehicles, engineering, naval systems, armament technology including explosives research, terrain research, works study, system analysis and life sciences including high-altitude agriculture, physiology, food technology and nuclear medicine. Projects are taken up in all these disciplines either in response to the expressed requirements of services (staff projects) or in anticipation of a future need for national defence (technology development projects).

The DRDO has been successful in various projects, Its recent achievements include the successful flight-testing of missile systems under Integrated Guided Missile Development Programme, namely, surface-to-surface tactical battlefield missile 'PRITHVI', medium range surface-to-air missile 'AKASH', third generation anti-tank missile 'NAG' and re-entry technology demonstrator AGNI. In addition, number of development flight trials of a short-range quick reaction surface-to-air missile "TRISHUL" have been conducted. A sophisticated low-level warning radar INDRA has been successfully developed and inducted into the services. An indigenously designed full-automated factory was set up by the DRDO to manufacture Fin-Stabilised Armour Piercing Discarding Sabot (FSAPDS). The other notable achievements of DRDO have been the multi-barrel rocket system 'PINAKA', India field gun MK-I and II, main battle tank ARJUN, bridge layer-tank, KARTIK, flight simulators for aircraft, air launched rocket powered missile target, 58 mm reusable rocket pod, field artillery radar, high technology EW systems, communications systems, night vision devices, new family of light weight Indian small arms systems (INSAS), cluster weapon systems, 81 mm illuminating ammunition for enhancing night fighting capabilities of Army, fuel air explosives for clearing path in the mine field, advanced material-composites for defence applications, advanced sonars, sonobuoys, naval decoys and simulators, rockets for ship defence, etc.

In addition, several high technology projects which include light combat aircraft (LCA), gas turbine engine, pilotless target aircraft (PTA), are at various stages of development in DRDO. The successful completion of these projects would greatly contribute towards self-reliance in critical areas of defence technologies.

Development of sea mines for use against submarines and ships is in an advanced stage. Upgraded version of the mine for underwater work by naval divers and frogmen has been developed and has undergone trials by the Navy.

A defence technology park is being set up in Bangalore to facilitate transfer of technologies by DRDO laboratories to industries. This will result in indigenous production of high technology items needed by the armed forces.

Research in health is being done in the areas of radio-immunoassay, radiobiology and bioengineering at the Institute of Nuclear Medicine and Allied Sciences, while the Defence Institute of Physiology and Allied Sciences carried out studies on high altitude acclimatisation and devises methods for increased fighting efficiency, safety and comfort of the armed forces.

The Defence Food Laboratory and the Central Food and Technological Research Institute of CSIR, Mysore have developed products, processes, preservatives and suitable packaging materials so that suitable Indian packaged food is provided to the armed forces in difficult terrain.

India is successfully tested various missiles:

TRISHUL is a short-range surface-to-air missile (SAM) with solid composite propellant and dual thrust motor and command to line-of-sight guidance. It has radar guided weapons and can be used against aircraft and helicopters. It is being developed for use of the army, navy and air force. The naval version has sea-skimming capability. The Trishul combat vehicle from which the missile is launched is an autonomous tracked vehicle with surveillance and tracking radars and missile guidance system for aircraft interception capability at very low levels.

AKASH is a medium-range (about 25 km) multi-target missile — it can target four to five enemy aircraft and missiles at a time. It has integrated solid propellant stages with command guidance followed by active homing. It has phased array radar for multi-target acquisition, tracking and guidance. It will be the first Indian missile system to use the ramjet principle for its propulsion (atmospheric air is used as one of the agents of combustion). The radar — Rajendra — is capable of tracking many targets simultaneously. The technology is said to be close to the system of the American Patriot missiles.

PRITHVI is a surface-to-surface missile with a twin gimballed engine using liquid propellant. It has strap down inertial navigation and is controlled and guided all the way by computer on board. While the short-range version is for the Indian Army, a longer range variant is being developed for the Indian Air Force. The Prithvi can go up to a maximum of 250 km with a 500 kg warhead (or payload) or 150 km with a 1000 kg warhead. The missile is fired from an indigenously developed tetra-mobile launcher. Prithvi is in the same class as the Soviet Scud missile except that the former's flight is computer controlled throughout while the latter follows a purely ballistic trajectory. It would be difficult to spot the Prithvi on radar or trace its trajectory and impact point because of its supersonic speed and limited flight time.

NAG is an anti-tank missile (ATM) with smokeless, high energy solid propulsion and, for guidance, a millimetric wave (MMW) radar seeker as well as an imaging infrared (IIR) system. With a range of 4 km, the missile has 'fire and forget' capability; the target sighted, the missile automatically aligns itself to it. Launched from a tracked carrier for a helicopter it is capable of piercing the most sophisticated of tank armours at the top.

AGNI is officially termed a 're-entry technology demonstrator', but for all purposes it can be considered an intermediate range ballistic system. It consists of a solid propellant first stage and a liquid propellant second stage. While the missile is hurled into space by a solid fuel booster, the liquid-fuelled second stage leads the missile warhead to its predesignated target. It has a range of 1000 to 2500 km and a take-off weight of 19 tonnes and a payload capacity of 1 tonne. Agni is capable of carrying a multipurpose payload. Indigenously built, one of its unique features is the heat shield of the re-tendry vehicle. Made of a carbon-carbon composite, the heat shield is able to withstand temperatures of 5000 degrees Celsius. The hear shield being ablative, it melts and chars and itself outer layers peel off, taking away the heat layer by layer.

The Pinaka

The Pinaka multi-barrel rocket launcher weapon system has undergone development trials successfully and is ready for user-trials by the Army before it is inducted into the arsenal of the service.

The Pinaka is a mobile system with 12 solid propellant rockets capable of delivering 'saturation fire' over targets that cannot normally be engated by artillery guns. With a range of 39 km, the Pinaka system can launch a variety of war-heads; pre-fragmented, submunition, and incendiary. Designed and developed by the Armament Research & Development Establishment, Pune on the basis of qualitative requirements formulated by the Army, the system has a quick reaction time, high accuracy, and excellent mobile characteristics. Comprising a launcher, rockets, replenishment-cum-loader vehicle and a command-post vehicle, Pinaka is said to be contemporary with other systems of its class that have been developed or are being developed, anywhere in the world.

Radar Systems

Radar, for Radio Detection and Ranging is a device for detecting and locating stationary or moving objects. A typical radar uses a powerful radio transmitter that detects objects with its radiowaves and detects the reflected waves with the help of a radio receiver. But also being produced now are ultra-wideband radars that make use of electro-optic computer and the acousticoptic process to give more data on the detected objects. Technological features such as moving target indication, track-while-scan, pulse doppler, frequency agility and displays are being incorporated in many radar systems. The smarter systems' also possess computers and advanced signal and data processing techniques.

Based on their functions, the radar systems can be grouped as land-based air defence systems, battlefield and ground surveillance radars, airborne surveillance radars, airborne fire control radars, naval surveillance and navigational radars etc.

In India, the laboratories of the DRDO, Electronic and Radar Development Establishment (ERDE), Bangalore and the Defence Electronics Research Development Laboratory, Hyderabad, have been working on developing indigenous radar systems for defence. Many of the indigenous developments have also been successfully commercialised. Bharat Electronics, public sector undertaking the Defence Ministry, has played a leading role in this field. Hyderabad-based HAL, also a public sector company under the Defence Ministry has also been involved in developing ground control radar. IFF systems (secondary surveillance radars) and other air-borne systems.

Bharat Electronics started radar production in the 1960s. The super-fiedermaus (SFM) radars were manufactured for the army. In-house R&D labs were set up at the same time. A storm-warning radar was developed and frigate radars, that is, medium and long-range surveillance radars in L&S bands and fire control radars (VM44/45) and display systems (DS 22) were productionised. Marine navigational radars were also productionised. In-house efforts have resulted in S&L band frigate radars, meteorological radars such as wind finding and cyclone warning radars, and MET radar for artillery (ARTYMET). A plant established in the mid-1970s by Bharat Electronics mainly for radar productionisation has produced the 3D Mobile (PSM33), 3D static (IRMA), Low Flying Detection Radar (INDRA) etc. A second generation of navigational radars, RANI and RASHMI, has been developed in the nineties. Efforts have also resulted in a surveillance and SSM control radars (APARNA) and a naval fire control radar RAWS-03 which are undergoing trials and the Precision Coherent Monopulse C-Band Radar (PCMC) for tracking satellites and other high speed objects.

India's indigenously designed and developed main battle tank (MBT), Arjun, was thought of in 1974 and was entrusted for development to the DRDO and the Combat Vehicle Research and Development Establishment (CVRDE), Avadi. In 1993, the DRDO offered six Arjun tanks of Mark I series to the Army for user trials, in January 1996 the MBT was dedicated to the nation.

The 58.5 tonne tank with its state-of-the-art technology, superior fire power, mobility and weapon systems is rated among the top three MBTs in the world. It has a top speed of 70 km per hour. Although it is slightly heavier than the Russian-built T-72s now in service in the Indian Army, Arjun exerts a low ground pressure for easy mobility in difficult terrain. With its 600 hp diesel power pack, Arjun can climb a gradient of 35 degrees for employment in the Rajasthan sand dunes. It is also designed to overcome the 'ditch cum bund' defence often encountered in riverine tracts. It has also a sophisticated weapon system with night vision facility. Its 120 mm gun with the newly developed super velocity finstabilised armour-piercing discarding sabot (FSAPDS) ammunition has formidable fire power.

Arjun has an armour plating of composites called 'Kanchan' successfully developed in India which is able to resist anti-tank missiles.

The indigenously developed Light Combat Aircraft (LCA) is likely to replace the MiG-21B1S fighters scheduled to be phased out by 2015. The LCA is a part of the Plan 2005, a decade long project of the DRDO for meeting 70% of military hardware requirements with national resources. Two prototypes of the LCA have been sanctioned for the project, the completion of which has been advanced to 2002 instead of 2005.

The LCA is in the same class as the advanced F-16 of the US and the JS-39 of Sweden. The technology demonstrator will have 70 per cent indigenisation. The LCA is to be a fly-by-wire aircraft with technology from the US firms Martin Marietta and Sunstrand. The first technology demonstrator will fly on imported General Electric F-404 engines, but the subsequent ones will have Kaveri engines which the Gas Turbine Research Establishment is designing and developing. The Kaveri engine is to have advanced control systems to make sure that the thrust level does not fall with the rise in ambient temperatures. The LCA is to be a multi-role modern aircraft with capabilities for fighting, intercepting and lombing. The air frame of carbon composites and aluminium alloys is a spin-off from ISROs research on PSLV.

The LCA uses an unstable configuration. Conventionally, aircrafts use a stable configuration. In an unstable design, the stability is duplicated by onboard computers which continuously adjust the aircraft's control surfaces suitably. But its latent instability gives these aircraft greater manoeuvrability, which is an enormous advantage to a fighter.

The LCA is said to have the smallest radar cross-section of any non-stealth fighter aircraft in the world. It carries a full suite of electronic counter measures (ECMs) and electronic counter countermeasures (ECCMs), all of them indigeneous, HAL, Hyderabad and the Bangalore-based Electronic & Radar Development Establishment (ERDE) are jointly developing the LCA's multimode radar.

The LCA was designed from the beginning to carry Indian Russian and Western missiles. It can handle beyond the visual range (BVR) missiles.

The fourth and final prototype of the military version of the Advanced Light Helicopter (ALH) has been tested. The LAH is a twin-engined cost effective multi-purpose and multi-role helicopters with rugged design to meet the stringent requirements of the armed forces. Designed and developed by the Hindustan Aeronautics Limited, the helicopter is in the four to five tonne class. It incorporates state-of-the-art technology to meet the diverse operational requirements of the Air Force, the Navy and the Army as well as the civilian market. It has made good use of composites with the resultant advantages of high strength and stiffness per unit weight, resistance to corrosion, high tolerance of damage and high fatigue life.

Nishant is India's indigenous remote piloted vehicle (RPV). Earlier called the Falcon, Nishant is soon to be inducted in the army. The Nishant, intended for battlefield surveillance and reconnaissance roles, incorporates advanced design features comparable or superior to those developed elsewhere in the world. Capable of carrying a 45-kg payload, it could cruise at about 150 kmph and fly for more than five hours. It could be controlled from the ground for distances up to 160 km, but is also programmed for autonomous flight. Made entirely of fibre reinforced glass, it is difficult to detect on radar. Its radio links to the ground are difficult to jam since they use the spread spectrum technique.

The Defence Scientific Information and Documentation Centre (DESIDOC) is the central agency for collection, processing and dissemination of scientific information.

Much of what is needed by the services is now being developed and produced indigenously in the several ordnance factories and eight defence public sector undertakings for defence production.

The ordnance factories operate a wide range of technologies and product-mix. Some important products are field guns, anti-aircraft guns, mortars, various small arms, ammunition for weapons and guns, rockets, projectiles, pyro-technics, bombs, grenades, mines, demolition charges, depth charges, infantry combat vehicles, battle tanks, self-propelled guns, transport trucks and patrol vehicles, high-altitude and combat clothing, optical and fire control instruments, engineering equipment, supply dropping equipment including parachutes and parasails, rubberised items, various knitted and woven items like blankets, web equipment and a wide range of general stores. The country is nearly self-reliant in ammunition, explosives and small arms. In the area of artillery, indigenous design of Indian field gun has been developed and brought into production. Ordnance factories are also successfully producing bombs required by the Air Force. Liquid fuel required for successful flight trials of PRITHVI and AGNI missiles, igniters and copper liners for the warheads were also developed.

The production of AJEYA Tank (T. 72) and SARATH (BMP-II) infantry combat vehicle has started at Heavy Vehicle Factory, Avadi and Ordnance Factory, Medak respectively. The requirement of engines for these two vehicles is planned to be met from the production in engine factory at Avadi. A modern and sophisticated opto-electronic production unit at Dehradun is to produce the opto-electronic fire control and night vision devices required for the tank and combat vehicle. Completely indigenous production of AJEYA Tank and SARATH Infantry Combat Vehicle is planned in 1994-95.

Ordnance factories have gone in full-swing in the field of passive night instruments. Three types of instruments have already been supplied and orders are being executed for another two types. A breakthrough has also been achieved in the development of specially ruggedised fibre optics cable.

Eight public sector undertakings under the Department of Defence Production and Supplies have the capacity for design, development and production of a wide range of equipments and components to meet the requirements of the defence forces. Incidentally they also cater to the needs of the civil sector.

Hindustan Aeronautics Ltd. (HAL) designs, manufactures, repairs and overhauls various types of aircraft, helicopters and related aero-engines, instruments and accessories. HAL at present manufactures Jaguar, Dornier-228 and MiG-27M aircrafts, Cheetah and Chetak helicopters and many types of aero-engines. HAL is also working on the design and development of an Advanced Light Helicopter, the first prototype of which has been successfully flown. The company is also participating in the design and development of the Light Combat Aircraft.

Bharat Electronics Limited (BEL) started as a single unit at Jalahalli, Bangalore. Since then, it has emerged as a leader in professional electronics, and has nine units at Bangalore, Madras, Hyderabad, Ghaziabad, Pune, Machilipatnam, Taloja, Panchkula and Kotdwara. The items manufactured by the company include low and high-power communication equipment in the HF, UHF, VHF and microwave ranges, high-power static and mobile radars, static and mobile troposcatters with line of sight equipment, broadcast transmitters, gun control equipment, weapons control system for frigates and electronic voting machines. The components produced include X-ray tubes, black and white TV picture tubes, image converter tubes, germanium and silicon semi-conductors, intergrated circuits and magnesium manganese dioxide batteries.

Bharat Earth Movers Ltd. (BEML) manufactures heavy earth moving equipment such as excavators, bulldozers, cranes, etc, sophisticated special assemblies like planetary axies, power shift transmissions and control valves and integral coaches for Railways at its three production units in Bangalore, Kolar Gold Fields and Mysore. The company is setting up facilities at Mysore to manufacture diesel engines in the range of 90 to 1000 horse power for captive use on its earthmoving equipment

Mazagon Dock Ltd., (MDL) Bombay, Goa Shipyard Ltd., (GSL) Goa and Garden Reach Shipbuilders and Engineers Ltd., Calcutta, are leading ship-building and ship-repairing units in the defence public sector, having facilities to build sophisticated warships like frigates, seaward defence boats, survey vessels, offshore and onshore patrol vessels. They also build merchant ships. MDL has constructed three indigenously designed Godavari class frigates. GSL has developed from a ship-repairing unit a medium sized shipyard on west coast, building seaward defence boats, landing craft and survey craft for the Navy.

Bharat Dynamics Limited at Hyderabad has the prime objective of establishing a production base for guided missiles and for the development of missile production technology in the country. The company undertook the manufacture of SSHBI, the first generation anti-tank missile, under licence agreement with Aerospatiale of France. The indigenisation achieved in the manufacture was to the extent of about 73 per cent. The company commenced the manufacture of a second generation anti-tank missile, namely, Milan in 1985. The progressive increase in the indigenisation of this missile system is still continuing. The company has diversified and produced rockets and other electronic systems for the Navy and the Army.

Mishra Dhatu Nigam Limited (MIDHANI), Hyderabad, has created indigenous capabilities for the manufacture of special metals and super alloys required for strategic, sophisticated industries like aeronautics etc.

The Defence Research and Development Organisation's research and development activities for military applications continues to generate spin-off technologies for civilian use. The new technologies have benefited various sectors such as metallurgy, agriculture, avalanche forecasting, and medical services which do not find it economically feasible to develop these technologies independently.

Recently, the DRDO developed the world's first water cannon. Equipped with automatic pulsating jets, the cannon would be useful in riot control.

Technology for 'modified atmosphere' storage of potato has been disseminated among farmers in Leh so that potato requirements in winters can be met by local produce. Appropriate agro-animal technologies have been extended to farmers in Ladakh and also, the Nubra valley/stachen glacier region. A multi-purpose gamma irradiation facility has been set up to sterlise medical and bio-fertiliser products. A plant process for CR, a new generation tear gas, has already made available a limited amount of the gas to the BSF for field trials. Avalanche forecasting work has helped the DRDO periodically issue avalanche warnings to susceptible areas.

For the spin-off of defence technologies to the medical sector, a society for Bio Medical Technology has been set up that includes the DRDO, the Department of Science and Technology. Ministry of Social Welfare, Department of Rural Development, Nizam Institute of Medical Sciences, Hyderabad and LV Prasad Eye Institute, Hyderabad.

9

LASERS

Lasers, Associated by many with the Strategic Defence Initiative (SDI), popularly known as Star Wars, are being extensively used in India for peaceful purposes.

The laser beams which are a coherent beam of light intense enough to vaporise the hardest material ever known, are being used to drill holes in diamond, to weld detatched retina of eye, and destroy malignant brain tumours located deep inside the brain or in the spinal cord and to transmit information.

Over 90 research institutions, universities and a few private companies in the country are working on various aspects of lasers, like their fabrication, development of laser materials and their applications to different fields like metrology, communications, medicine and isotope separation, and in basic research.

Work on fabrication of lasers has been taken up at a number of institutions in this country and they have successfully fabricated nitrogen, carbon dioxide and helium-neon lasers, Scientists believe that indigenous commercial production of nitrogen, argon, helium, cadmium and nitrogen-pumped dye lasers (all by Central Electronic Ltd., Sahibabad), carbon dioxide laser by Advani Oerlikon and Joyti Ltd., Helium-neon by Electronics Corporation of India Ltd., may soon become popular and reliable.

Attempts to develop lasers in this country started as early as 1964. The activity could not generate enough momentum in the first decade, but has started picking up since the mid-seventies. Though the total efforts are still at a modest level compared to other advanced countries, many institutions and research organisations have taken up the know-how or started their own R & D activity for production and marketing of laser.

While the argon laser prototype was developed by IIT, Kanpur, Banaras Hindu University, Varanasi, made a pulsed carbon dioxide laser for work on photo-chemistry.

Due to their unique properties, lasers find application in various fields. Some of the important applications of lasers are being discussed here.

The revolution in the field of spectroscopy was made by lasers due to the availability of intense and narrow line width radiation. In the measurement of the absorption spectrum these are advantageous in detecting weak absorption lines. There are several techniques for detection of weak absorption spectra like optoacoustic spectroscopy, multiphoton ionisation spectroscopy, etc. There are also non-linear spectroscopic techniques like two-photon absorption spectroscopy. The laser has reduced the time required to record a Raman Spectrum from hours to minutes. The laser most widely used is argon ion laser for Raman Spectroscopy.

In chemistry the lasers find use both as a diagnostic tool and as a means for inducing chemical reactions. It also helps in separation of isotopes of an element. Lasers generally used in chemical applications are dye laser, excimer lasers. CO_2 lasers and Nd: YAG Lasers.

Industry lasers are finding great applicability in industry also. At present lasers are routinely used for surveying for material processing like welding, cutting, glazing, alloying cladding, for non-destructive testing etc. It is used in aligning structures like bridges, large buildings, tunnels, pipes, mines etc. Lasers have completely replaced conventional drilling techniques to drill holes in diamonds, for making are drawing dies, gems, turbine blades of jet engines; they are widely used for cutting metals, ceramics, plastics, cardboard, cloth etc.

Laser annealing of solar cells improves their performance. Lasers can be used to write directly on silicon wafers, the patterns required to make ICS.

Lasers are widely used for making on plastic, metal products etc. Laser scanners are used to read bar codes on consumer goods in shops, to inspect precision components in workshops, to identify goods wagons in a moving train and to read text and other documents. Laser printers are fast and of high quality. Lasers are used for heat treatment of surfaces.

In modern warfare the lasers find great use in weapon systems for finding tanks and artillery gun, measure accurately the range of targets, thus giving first hit capability.

Ye-Ne laser help in simulating the firing of rifle guns while training soldiers and also improve the aim of guns.

Laser isotopic separation (LID) promises to be the most economical way of enriching natural uranium.

Lasers are also playing a crucial role in the search for a viable way of trapping fusion power.

Lasers are finding use in almost all areas of medicine. The highly powered laser beam with high intensity is adequate for cutting tissue and thus a great surgical tool which can replace the traditional scalpel. Advantages include : (i) prevents bleeding (ii) lowers possibility of infection, (iii) does less damage to the nearby cells.

Lasers are routinely used to wed the detached retina to the chorold. Other serious conditions of eye which are treated are diabetic rehonoputhy, macular degeneration and haemorrhage. For the treatment of these diseases the Argon or Krypton ion laser is used. Glaucoma is a disease of eye in which the pressure inside the eyeball increases, resulting in damage to the retina and ultimate blindness. To overcome this pressure and save the sight a small hole is drilled in the eye with Nd-YAG laser. Lasers are also used to correct the shape of the lens e.g. myopia through a procedure called radial keretotomy.

The precision of laser surgery has been a major advantage for delicate operations like heart-by pass surgery and neurosurgerical intervention. By using fibre optic endosope with a laser, it is now possible to cauterise the bleeding ulcer without making any opening in the body. Another example is the treatment of blockage of arteries supplying blood to heart, a condition resulting in the myocardial infection. The remedy so far was to shunt the blocked artery with one taken from another part of the body, a procedure known as by-pass surgery. Photo radiation therapy is used for treatment of cancer. Hemato-porphyrin derivative (Hpd) a dye has the property of attacking selectively the cancerous cells. When intense laser beam from gold vapour laser is irradiated, the Hpd molecule decomposes releasing singlet oxygen which kills the cancerous cells and tissue.

Lasers are used in the treatment of gall bladder and kidney stone effectively.

The main drawback of the entire line of light communication was their susceptibility to weather additions. Corning in 1974 produced low loss optical fibre from high purity materials. Now there is remarkable progress in both optical fibre and semiconductor laser technologies. The loss of optical fibres has been reduced to less than 0.3 db/km at 1.5 micron wavelength (i.e., a 10 km long fibre will transmit 50% of incident light).

In addition to transmission of data, optical information processing and optical computers are related issues. Optical information processing is used for fingerprint identification, processing of photographs taken by satellites and high flying aircrafts etc. Optical computers, based on optical bistable device will offer a manyfold increase in the speed of computing and also the capability to do parallel computing.

Data storage is another area where higher density of storage is possible by using optical methods. The storage medium is generally a thin film of metal whose optical property such as reflectivity gets modified when illuminated with a powerful "WRITE" Laser. The lower power "READ" laser reads the change in optical property as the information required. It takes less than one square micron to record one bit of information. Laser video discs are widely used as a source of entertainment. Though the laser optical data storage disc possesses high storage capacity over the magnetic disc yet it is not possible to erase the information written from optical data disc. In a compact disc (CD) laser light is used to read instead of a needle, so the grooves can be made very small (even ten millionth of a meter). Exceptional fidelity is achieved as amount of information stored can be very large. A semiconductor laser reads the CD by bouncing off light from the disc and electronically processing it. CD ROMS (Read only Memories) enable us to store even encyclopaedias on single discs. Conventional computer memory devices work on basis of magnetic recording and reading of data, but optical discs have the advantages of larger storage durability and faster access for retrieval of data.

The importance of laser was well recognised by the scientific community in India at various institutions as early as the mid-sixties. In 1964 at BARC first gallium arsenide semi-conductor laser was fabricated. This laser was used in 1965-66 to establish an optical communication link between BARC and TIFR. The laboratory with the largest effort for developing laser technology in India is BARC. BARC has developed a 50 MW laser for Raman Spectroscopy. It has also worked for development of various CO_2 Lasers. BARC has also developed the solid state lasers namely Nd: YAG, ruby laser, Na glass laser. The Centre for Advanced Technology (CAT) has planned an extensive programme on laser development and engineering. It has also proposed to undertake limited scale production of laser and laser instruments. CAT has taken up production of 10W CVL.

DRDO has developed laser range finders for tanks which use switched Nd: glass laser. DRDO is developing laser material and semiconductor lasers. It has excellent facilities for growing crystals by several techniques and has been successful in growing Nd: YAG crystals and Ca As crystals.

Various IITs, particularly those at Kanpur and Madras, are working in the area of laser development, as are the NPL and the Indian Institute of Science.

India's research base is good, but the research in laser has not been adequately exploited to technological and commercial advantage by policy initiatives till recently. Some efforts has been made to rectify the situation through formulating the National Laser Programme, a joint strategy of the Departments of Science and Technology, Atomic Energy and Electronics under the Eighth Plan. The National Laser Programme, aimed at developing and manufacturing important lasers for various applications will soon have indigenous lasers and laser-based equipment at much lower cost which will, substitute import of laser equipment, laser-producing crystals and related equipment. The Crystal Growth Centre of Anna University, Madras, has been selected to produce laser-producing crystals.

The Central Scientific Insturments Organisation (CSIO), which has been working on holographic technology for 14 years, has developed a variety of holograms for different applications.

Holographic technology is the science of creating three-dimensional images that are perfect substitutes of the original. Unlike the conventional photograph, the hologram, created with the help of laser light, is a faithful and complete record of the original object in three dimensions. It looks like an ordinary pieces of glass, but when light is turned on the image, an exact replica of the original comes into full view. The potential beneficiaries of the technology are jewellers, curio-manufacturers, museums, security and advertising agencies. Jewellers need not risk theft by displaying original ornament on their shop windows. They can substitute them with holograms.

On the whole, India needs to enhance the use of lasers in the cause of development and human welfare. It is sad that only a handful of hospitals, for example, make use of laser surgery.

10

ELECTRONICS AND INSTRUMENTATION

India has now expertise to manufacture modern electronics and precision instruments both for research and routine application in diverse fields like nuclear energy, space and defence.

While scientists of the Bhabha Atomic Research Centre in Bombay have developed a number of control and instrumentation system for the nuclear reactors like Dhruva and Purnima II, over half a dozen Defence Research and Development Laboratories have designed and developed modern electronic and precision instruments — "eyes and ears of the defence system."

The BARC Scientists have also designed and developed a number of sophisticated special purpose instruments for the Reactor Research Centre at Kalpakkam, Nuclear Power Stations, Fuel Reprocessing Plant, Radiation Medicine Centre, Indian Institute of Astrophysics and Defence Electronics Applications Laboratory.

A wide variety of sophisticated equipment, useful for research as well as for routine applications in diverse fields, have also been made by BARC scientists. Important among them are the TDC-312 computer based gamma scanner, high efficiency solar cells, the microcomputer based metallographics analysers, the computerised plant data acquisiton system, microprocessor based nuclear instruments like liquid scintillation spectrometers, multichannel analysers, mossbauer and auger electron spectrometers, field ion microscopes, quadruple mass analysers, high power electron beam furnaces, plasma torches and infrared sensitive image converter tubes, various pieces of equipment like microprocessor based thermoluminiscent and film badge dosimeters, radiation and contamination monitors, wide range survey meters and infrared gas analysers have been developed by BARC for personnel and environment monitoring.

Besides, BARC has also set up the Reliability Evaluation Laboratory which offers services for testing and measuring electronics components.

The history of the Indian computer industry dates back to 1955 which was marked by the dominance of IBM and ICL. While Electronic Corporation of India Ltd (ECIL) was formed in 1967, the foreign exchange crisis of the 1970s caused tight control on imports. The new FERA regulations in 1976 caused the exit of IBM and the dilution of ICL to form ICIM. The period from 1984 onwards saw some liberalisation measures initiated by the Rajiv Gandhi government. The growth of the industry in the post-1984 era was impressive enough. The introduction of the Personal Computer (PC) provided a boost to the IT industry in India. While the percentage growth of the computer industry worldwide was down to single digit figures, in India it was still experiencing a 40 per cent plus annual growth. Even in exports, both hardware and software experienced healthy growth.

Over the years computers have become faster and more powerful. As a consequence, various branches of science now make use of the computer's growing capabilities for modelling and simulation studies. Integrated circuits are designed and tested wholly on the computer. Computer simulation studies help decide the most suitable design for variety of structures, from vehicles to high-rise buildings. Molecular modelling helps chemists gain a clear idea about the properties of substances and thus develop specific materials. There are problems such as monsoon models, gas dynamics, molecular biophysics, quantum chemistry and reactor dynamics that cannot be simulated on the conventional computer. The need for high-speed numerical computing is felt as the number of computations to be performed in many problems is non-linearly related to the number of data points or dimensions. If the matrix size doubles, the number of operations will increase eight-fold.

Serial machines have changed little since mathematician John von Neumann defined the principle in the mid-1940s. But the general-purpose computer he helped create does one job at a time.

And though processor chips now work phenomenally fast, there is a physical limitation — known as the von Neumann barrier— beyond which higher speeds do not lead to faster turnarounds.

Currently, Supercomputers such as ETA, CRAY and NEC's SX2 are used to solve such problems. They typically achieve their speed by using a method analogous to an assembly line operation. An operation is broken into a set of sequential sub-operations. Each sub-operation is performed by independent hardware circuits. If a set of operand pairs are fed to the pipeline, then one result per operand pair will be available at a time interval equal to one sub-operation. For example, if a multiplication operation takes 1 microsecond and if it can be broken into 100 subtasks of 1/100 microseconds each, and 100 operand pairs are fed to the assembly pipe, one product will appear out of the assembly pipe every 1/100 microseconds. The method will work only if there are enough operand pairs to fill the pipe. The ratings such as 500 million arthimetic operations, quoted by supercomputer manufacturers, refer to peak ratings obtainable when all the pipes are filled. Average performance is found to be only about 10 to 15% of peak performance. Further, the design philosophy of supercomputers, with a primarily serial architecture, and achieving high speed by using special super speed arithmetic circuits and the assembly line approach, is very expensive and reaching saturation.

An alternative approach currently being widely explored is to use a large number of less expensive processors working simultaneously on a task which is broken down into a large number of subtasks where each subtask can be carried out independently and at its own pace. Each processor is quite powerful and can perform a variety of tasks. This is to be contrasted with pipeline processing in which a sequence of identical tasks are fed to a pipe. The organisation of a large number of processors which work cooperatively is called a parallel computer system. The basic challenge in designing a parallel computer system is in solving the problem of coordinating the tasks of a large number of independent computers without wasting time and effort in unnecessarily passing data and instructions among them. A parallel computer can be scaled up tremendously for higher performance.

Computer speeds are measured in MIPS or million instructions per second — the number of operations the processor at the heart of the computer can perform. Parallel computers like to judge their performance by another yardstick, Mflops of Mega Flops—short for one million floating point operations a second. Floating point is computer shorthand for storing very large numbers. Claims by different parallel computer designers for speeds obtained, are difficult to compare unless they are achieved by solving identical problems under identical conditions — which is rarely possible. For a relative comparison of speed, a benchmark program called LINPACK is run; it permits the measurement of floating point performance of computers using a standard system of algebraic equations.

Artificial intelligence (AI) is the study of ways in which machines can be made to have sufficient creative reasoning power to perform mental tasks at which, at present, human beings are better. Examples of problems that fall under the area of AI include common sense tasks such as understanding a language, recognising scenes, finding a way to reach an object that is far overhead, heavy and fragile, and making sense of the plot of a novel. In addition, AI includes expert tasks, such as diagnosing diseases, designing computer systems, planning scientific expeditions.

A machine with true artificial intelligence has not been created. AI studies have been focusing too narrowly on 'problem solving' as a paradigm for all intelligent behaviour. This has led to a paradoxical situation where AI specialists have been able to construct sophisticated knowledge based expert systems, but are unable to come to grips with ordinary everyday perceptual motor behaviour involving vision, speech, locomotion, manipulation, language, etc. Computers at present are smart in some ways but are not intelligent. They 'mechanically' work out solutions to problems; they do not use a logical/intuitive approach that characterises humans.

Expert systems are the bridge between human and computer methods of problem solving. Through the application of AI techniques, expert systems capture the basic knowledge that allows a human to act as an expert when dealing with real life problems. One of the foremost requirements for an AI technique is that it must copy/reflect as closely as possible the human judgement/reasoning and intuition used in solving real world problems which are characterised by uncertainty.

One of the reasons for the computer's failure to accurately mimic human reasoning can be the use of conventional and classical bivalent/binary logic. In bivalent/binary logic, every event has only two states—either it occurs or it does not. Obviously, such a representation is inadequate for modelling the understandably complex real-life situations as some events may run to partial fulfilment and yet be important enough. Fuzzy reasoning is specifically designed to deal with the inexactness (or fuzziness) that is present in the knowledge used by human experts.

The world 'robot' comes from the Czech robota, meaning compulsory service. Robotics is the study of the design and use of robots or machines programmed to carry out a series of operations without human guidance. Robots fulfil an important role in industry; they do simple and repetitive jobs more steadily, more reliably and more uncomplainingly than a human being could or should, as Isaac Asimov said. Robots are a means of maximising productivity. Manufacturing technology, a key factor in attaining high levels of productivity, is today going through, what may be called, the third Industrial Revolution, benefiting from the rapid progress achieved in the areas of computers, robotics, expert systems, intelligent sensors and controls, and information and communication technology.

There are two major types of automated manufacturing; flow-line manufacturing and batch manufacturing. In flow-line manufacturing, dedicated capital-intensive equipment is designed for high volume production of a single product. This has automated monitoring and control systems using distributed computer technology. High volume production of discrete components such as automobiles, two wheelers, television sets, microcomputer systems, aircraft, etc. using highly specialised dedicated equipment, popularly known as hard automation, has proved its effectiveness.

Batch production, on the other hand, represents a large proportion of manufacturing activity in metal cutting, electronics, plastics, ceramics, wood and garment industries. The automated manufacturing technology (AMT), using computer controlled machine tools and automated material handling equipment, and supervised by a central computer, offers much promise in batch production. This includes high quality, accuracy and repeatability of the product, increased machine utilisation, reduced lead times, reduced inventory, lower labour costs and quicker response to new product design.

Two streams of applications of information technology to manufacturing can be identified. The first comprises direct applications where the computer, or a network of computers, monitors and controls the hardware in a manufacturing plant, such as machine tools, inspection machines, robots and automated guide vehicles, which physically transform or transport of assemble components. Indirect applications include computer aided design (CAD), production planning and control, inventory control, marketing, sales and other factory management systems (FMS) used for management of production. Business software such as materials requirements planning (MRP-I), manufacturing resources planning (MRP-II), just-in-time techniques (JIT) and optimised production technology (OPT) is available for indirect applications. Automated manufacturing technology provides the hardware for direct applications and is highly capital intensive. Computer Integrated Manufacturing (CIM) includes both types of applications.

Industrial robots are being increasingly used in several engineering applications including manufacturing. Tasks such as welding, spray painting, machine loading, material handling and assembly constitute the major use of robots on the factory floor. Recent advances in machine vision and expert systems and their use in the development of intelligent robots have made robots versatile and integral parts of the automated factory. Indeed the number of robots used has become synonymous with high productivity. There is also a widespread use of robots in hazardous environments such as toxic chemical industries, assembly of explosives, deep sea, undergound, high altitude and mountainous terrains, cold climates and nuclear reactors.

Robots are being used on a rather low scale in India. An FMS in our country is installed at the Heavy Alloy Penetrator Project (HAPP) at Trichi. This system has intelligent CNC work centres, integrated with a tool management system and loading/unloading gantry, and is capable of uninterrupted operation for more than one shift. Material integration at the shop floor has been achieved by introducing automated guided vehicles, and information integration by a supervisory, computer, inspection through electronic gauge is integrated with the manufacturing process. Further, an integrated factory management information system is being adopted to interface with the automated manufacturing facilities.

Robots and other material handling equipment are being developed at Bhabha Atomic Research Centre (BARC) Bombay, for nuclear applications. The remote handling systems developed and deployed at various nuclear plants under the Department of Atomic Energy (DAE) have freed the country from dependence on foreign suppliers. R&D units of HMT, IITs, and IISc, CMTI are all working on developing robots or rather robot arms and sensors for various purposes. Advances have been made in computer numerical control (CNC) technology. The IISc has developed a robotic arm with microprocessor technology. A flexible turning cell, the FTC 25, developed by HMT loads and unloads items automatically. The BEL introduced a 'pick and place' type of robot in its production work. A full-fledged robot, able to move with ultrasonic sensors, has been developed at the Vikram Sarabhai Space Centre for robot hands with phallanges that are controlled by software and could be used for satellite recovery or other critical operations in space. Unlike conventional robots in which actuators are used for movement, in the VSSC model special alloys control the motion by changing shape with temperature changes.

A DRDO venture, the Centre for Artificial Intelligence and Robotics (CAIR) in Bangalore has developed 'Chaturobot', an 'intelligent' robot with vision sensors which can pick up objects in its visual field no matter how they are oriented. CAIR scientists have also perfected a new generation Expert System Shell, "Nipuna", which has two special features; completeness, that is, it never falls to find a conclusion that follows logically from the stated facts; and consistency checking, that is, Nipuna even looks for and detects inconsistencies in the data fed into it.

Light has for long been used as a medium for transfer of information — in the form of fires, lamps, light signals. But in the matter of long distance communication, light, as one perceives it, poses problems; it does not go round corners but corners need to be negotiated if distances are to be spanned. Furthermore, atmosphere as a medium for transmission of light would be unsuitable as dust, fog, rain, etc., cause heavy attenuation owing to absorption, dispersion and diffraction, limiting the range of even line-of-sight transmissions to just a few kilometres. This serious limitation called for a device which could entrap light within itself and, which could be flexed to negotiate physical obstacles. The other requirement was the availability of a source of light of high intensity. In 1970, a breakthrough was achieved when optical fibres with losses of the order of 20 decibels per kilometre were developed. The invention of lasers and light emitting diodes (LEDs) solved the problem of the light source. The continuing research activities finally gave birth to optical fibre based communication systems.

The velocity of light in various transparent media varies. A medium in which light velocity is lower is termed as denser than the other, in which the light travels faster. A light ray on crossing from a dense medium into a less dense medium, bends away from the normal of incidence. In one keeps on increasing the angle of incidence (i.e. the angle between the ray of light and the normal of incidence), then a stage shall be reached when the light, instead of being refracted, will undergo reflection. This phenomenon is known as 'total internal reflection'. If on such reflection, the reflected ray again meets a surface separating a transparent media identical to the one described above, then it will again undergo total internal reflection. Optical fibres are manufactured to satisfy this condition enabling light to remain trapped inside the fibre while it travels along the longitudinal length of the fibre.

The following constitute the basic building blocks of a fibre optic communication system: (a) a transmitter consisting of a light source and associated drive circuitry; (b) fibre encased in a cable to provide a practical method of spanning long distances with ease and safety; and (c) a receiver made up of detector, amplifier and other signal restoring circuitry.

Injection Laser Diodes (ILDs) and Light Emitting Diodes (LEDs) are the light sources used in optical fibre communication. These devices can be directly modulated by varying their input currents in consonance with the electrical signals to be transmitted. Alloys of gallium, aluminium and arsenide are used for light sources operating in 900 nm band. For wavelengths of 1100 nm to 1600 nm, alloys of indium, gallium, arsenide and phosphorus are used. Laser diodes are used for long distance communication since they produce coherent light beams that are highly monochromatic. Optical fibre is made up of a cylinder at the centre, i.e. at the 'core'. The fibre is a thin hair-like thread made of glass or plastic. The core is surrounded by a material layer, the 'clodding'. The core has a specific refractive index which is considered while making a choice for a particular application. Due to the differences in refractive indices between the core and the cladding, the light signal is trapped inside the core (due to total internal reflection) even if the wire is bent. There are two types of fibre optics depending on the size of the core — the single mode fibre (with core of 10 micron diameter) and the multimode fibre (core of about 50 micron diameter). In India, Department of Telecommunications (DoT) have standardised on 6 fibre, 12 fibre and 24 fibre cables while Indian Railways generally use 10 fibre cables. Composite optical fibre cables contain copper wires also for providing power to repeaters. Cables may be unarmoured or armoured with steel tape or steel wires. Different types of optical fibre cables are manufactured according to specific applications.

The photo-detector senses the luminescent power falling upon it and converts the power variation into corresponding varying electric currents. Semiconductor photo-detectors that are small in size and possess high sensitivity and fast response time are used. PIN photodiode and avalanche photodiode (APD) are the two types most generally used. Silicon, germanium and indium-gallium arsenide are the materials used in making photodiodes. Even though the fibre is extremely pure, there are minuscule impurities that absorb light and light may be scattered due to variations in composition and molecular density. Bending also causes a slight loss of energy. Also, light pulses overlap while travelling, causing difficulty in separating them at the receiver end. Despite all this, fibre optics communication has distinct advantages.

Optical fibre of communication systems provide the following major advantages over the other conventional systems:

(a) An extremely wide bandwidth, permitting several times larger volume of information than is possible through conventional means. It is possible to encapsulate several hundred fibres in one single cable though the DoT has so far not gone in for more than 24 fibres in one cable.

(b) Repeater distances of the order of 80 kms to 100 kms are easily possible today.

(c) The higher channel capacity, larger repeater distances and extremely low maintenance cost make the systems very cost effective.

(d) Optical fibre cables are 25 times lighter in weight and at least 10 times reduced in size as compared to conventional cables. This makes their transportation and installation much more easy. It also makes them attractive for airborne, shipborne and space applications.

(e) Optical fibre communications are free from radio frequency interference and are immune to electromagnetic interference and electromagnetic pulses. This results in unprecedented high fidelity.

(f) The communications are almost error-free.

(g) It is well nigh impossible to intercept of communication without being detected.

(h) The system requires simple maintenance. Very high maintainability and reliability are achieved.

(i) Power consumption is extremely low.

In India, the deployment of fibre optics in the public network began with the installation of an experimental junction link in Pune between two telephone exchanges, and then came a long trial route between Baroda and Ahmedabad. In the period 1985-88, imported fibre optic systems of about 2,500 km were installed. In 1988 came an indigenous effort when Optel Telecommunications Ltd. installed an optical fibre system for Bombay Port Trust linking Piprau with Butcher Islands.

Optel Telecommunications Limited, Bhopal and Hindustan Cables Limited (HCL), Naini are the two Indian Companies currently engaged in the manufacture of optical fibre and optical fibre cables. Optel launched regular commercial production of optical fibre cable in 1989. Optel also started regular commercial production of optical fibre.

Optel Telecommunications Limited and Indian Telephone Industries are the two Indian companies licensed for production of optical line terminals and other systems equipment.

Optel is the only Indian company having vertically integrated facilities for the manufacture of the entire range of products required for setting up of optical fibre based long distance telecommunications, with the capacity for undertaking design, engineering, installation, testing and commissioning of communication projects on a turnkey basis.

Optel and HCL began supporting the Department of Telecommunication (DoT) programme of nearly 4,000 to 5,000 route-km optical fibre systems per annum on an average from 1991. By March 1993, 9,960 route-km of optical fibre systems were installed and another 6,000 to 7,000 route-km is under installation. With manufacturing in this section being deregulated, several more optical fibre production units could be expected to come up.

The current annual demand of optical fibre-based communication systems works out to around 6,000-7,000 route-km (according to the Eighth Five Year Plan) and the required terminal equipment.

Telecom plays a dual role in the field of information technology. The first one is the management of information for its own operation — where it behaves like any other industry. The second role is that of a mass data transporter through its leased circuits, public switched telephone network (PSTN), packet switched public data network (PSPDN), satellite based remote area business message network (RABMN). Satellite communications, cable TV, cellular radio, and videotex are revolutionising the way we receive entertainment, keep up with the news and work, shop and even conduct financial transactions.

The Department of Telecommunications is implementing a very ambitious plan of computerisation of various facts of telecom operations and management. Since the early eighties, the department has inducted a large number of computer controlled electronic switched systems in its network. These switching systems in its network. These switching systems are complex real time computer machines designed to meet exceptionally high reliability and availability criteria. They are extremely fault-tolerant with on-line diagnostics and possibilities of system reconfiguration in case of faults. They contribute greatly to the quality of telecommunication services.

Information technology is now being extensively used to support telecommunication operations and management. Some of the activities include computerisation of billing, cable records, fault control, directory enquiry, directory printing, inventory control, store accounting and commercial waiting list. All these have resulted in a big improvement in customer services with prompt responses.

Another area where information technology has played a significant role in improving the operational efficiency of telecommunications is that of Network Planning with PC based tools for the optimisation of telecom networks.

The telecom mission also known as 'mission: Better communication' was divided into five mini-missions, aimed to improve the quality of service, increase accessibility to telephones, improve telegram delivery, provide telex on demand and improve rural telecom facilities. The time-bound ambitious programme has resulted in improvement in various services.

In the Eighth Plan DoT has taken a major initiative to provide a telephone in each of the 2.28 lakh gram panchayats by April 1995 and a Long Distance Public Telephone (LDPT) within a five km walk in every village by the year 2000. Considering that on April 1, 1991 only 40,000 out of 5.6 lakh villages had access to a telephone, it will be a stupendous task specially as resources with DoT are minimal.

Universal on-line information access requires data communications networks which are accessible to the public. For long only one was available in India, viz. the Public Switched Telephone Network in which data is transmitted through the normal telephone channels. A new system — packet switched public data network, which is more efficient and cheaper, has been created by DoT, viz. the I-Net. Packet switching may be described as "the routing of data in discrete quantities called packets, each in specific format and with a maximum size". Unlike the circuit switching used in telephone/telex networks, in this systems, circuits are not switched and dedicated to the users for the duration of a `call', but information to be exchanged between users is transmitted in packets which occupy circuits only during actual transmission of data in a cell. I-Net provides connectivity between computers and terminals at high speed. I-Net subscribers can connect with Remote Area Business Message Network (RABMN), international data networks and databases, and E-mail networks. I-Net offers closed user grouping, reverse charging, permanent virtual circuit and other facilities.

I-Net, now operational in Ahmedabad, Bangalore, Bombay, Calcutta, New Delhi, Madras, Hyderabad and Pune, will be extended to cover 49 cities by 1995-end and 89 cities by 1996-end. The existing facilities in the eight cities are also to be upgraded for greater capability.

International satellite communication became possible for India when the country became a member of the INTELSAT on commissioning its first earth station at Arvi, near Pune, in 1970. Domestic satellite communication in India commenced in 1977 using the Satellite Telecom Experimental Programme (STEP). Commercial service began in 1979 on capacity leased from INTELSAT; Delhi and Madras were linked to five remote areas like Leh in Himachal Pradesh, the capitals of Mizoram, Lakshadweep and Andaman and Nicobar Islands. With INSAT series of satellites being launched in 1983, the service truly expanded.

Satellite communication has advantage for a country like India with a large geographical area and a relatively poor terestial communication service. One, any signal transmitted from the satellite can be received by any number of locations within the area covered by the satellite. Two, points within the coverage area can be linked to one another without considering its location or communication infrastructure available at those points. A nationwide satellite communication network can thus be established in a short time as well as cost-effectively.

Very Small Aperture Terminals (VSAT) are earth stations with antennas less than two metre in diameter and they are useful in reducing the cost of terminals for large networks. VSATs are characterised by 'star' topology if they are interconnected only via a hub-station and not directly to one another as in 'mesh' system.

VSATs first came on the Indian scene in 1988 when the network of the National Informatics Centre, NICNET, was established. The network consists of a master (hub) earth station at New Delhi, connected to the host computer and to more than 650 VSATs covering all the districts of the country. The NIC is the decentralised depository of information for various central and state government departments for which large databases are being developed. NIC has four large mainframe computer system at Delhi, Pune, Hyderabad and Bhubaneshwar to maintain the databases, the Delhi system being the main host. Local Area Network (LAN) and minicomputers/super ATs installed in secretariats of the State capitals and Union

Territories are used to consolidate information on various districts and blocks. The LANs are being connected by a Metropolitan Area Network (MAN). PC-based terminals are expected to be installed at the block level in the third phase of the NIC plan. The LANs and MANs to be installed will be connected to a Wide Area Network (WAN), and this LAN-MAN-WAN will form the logical network structure that is to evolve over the next decade. The NICNET supports several closed user groups such as the Steel Authority of India Ltd. Border Roads Organisation, the Department of Biotechnology, etc.

A second VSAT network for public use came into being in July 1991, namely, the Remote Area Business Message Network (RABMN). Run by the DoT, its hub station is at Sikandrabad (Uttar Pradesh). It has an installed capacity to handle 1,000 VASATs, of which about 400 are being used. The hub capacity can be expanded to meet the need of more VSATs. RABMN can provide remote telex connection by linking the VSAT to the Telex Switch at Delhi. (RAVI)

NICNET as well as RABMN provide interactive data communications in a variety of forms with national as well as international connectivity via the Videsh Sanchar Nigam Ltd. (VSNL) gateway. Both networks have adopted the code division multiple access (CDMA) technology. In India this technology is licensed to ITI Equatorial Satcom Ltd., a joint venture company set up exclusively for local manufacture of VSATs.

Incidentally, CDMA is based on the Spread Spectrum Technology, whereby occupied radio frequency spectrum is deliberately spread by combining the data to be transmitted with a digital pseudo-random noise (PN) sequence or word. This word, unique to each VSAT, is used at the hub station to get the particular information from among the signals received, all simultaneously occupying the same frequency spectrum. CDMA technology enables one to reduce the size of the sources is negligible. Thus transportation and installation is easy, as is expansion of the network, since there is no need for additional frequency or channel allocation every time a VSAT is to be added.

The Videsh Sanchar Nigam Ltd. introduced the Gateway Internet Access Service (GIAS) in August, 1995. Thus Internet Services were made available to India on a commercial basis. The DoT and the Mahanagar Telephone Nigam Ltd. (MTNL) are also involved in the service. The service will enable users to access textual, audio and video information on virtually everything in the world and from any part of the globe connected through the Internet.

An integrated network service called 'SoftNet' has been designed and established in the Software Technology Park (STPs) for providing wide variety of value added services, such as Soft Point, SoftLink, SoftPac and video conferencing facility/distant learning. Video conferencing facility between Bangalore and the rest of the world has been established. STPs also promote new technologies such as software engineering and quality assurance.

INMARSAT, an international body set up in 1979, provides mobile satellite communication service — maritime mobile, aeronautical mobile and land mobile applications. India is a member of this organisation. INMARSAT operates four satellites — two over the Atlantic and one each over the Pacific Ocean and the Indian Ocean — in geostationary orbit. These provide a world service over the region between 86° North and South of the equator. INMARSAT provides a VSAT service in India. The VSNL commissioned a Land Earth Station at Arvi in 1992 to connect INMARSAT VSATs to public telecom work. For users in India VSNL provides billing service with payment in Indian rupees. The INMARSAT VSATs are the smallest and hence portable. The L-band of frequency (about 1500 MHz) is used between the VSAT and the satellite.

India has decided to join the global Iridium network. The Iridium system is a satellite-based wireless personal communications network designed to permit any type of telephone transmission — voice data, fax and paging —to anywhere in the world.

Some of the applications of VSATs available in India are as follows:

Text messaging and E-mail E.Mail is the facility for the despatch of documents in electronic form ranging from the ability to interconnect word processors to net-works where documents/messages may be stored and subsequently forwarded or retrieved. Besides reaching the destination very fast, electronic mail is cheaper than fax and telex. In this basic service text message can be sent in conversation mode or the store and forward mode of E-Mail. Both RABMN and

NICNET provide the facility. NICMAIL is an example of successful E-Mail service.

Facsimile VSATs use completely digital mode of transmission: hence every high quality fax service is possible between one VSAT and another. PC-based fax is quite common in India now.

Remote Telex RABMN is connected to the New Delhi telex exchange making it possible for its subscribers to have access to national and international telex networks.

International Packet Network Access. Both RABMN and NICNET provide connection to gateway packet switch system (GPSS) of the VSNL, allowing subscribers access to various international packet networks. Packet-switched networks facilitate smooth and highly economical transfer of data, for E-mail, electronic fund transfer, airline and hotel reservations, software development, etc.

Electronic Money Order. The Department of Posts is to implement a money transfer system using a VSAT network.

Voice Mail RABMN and NICNET VSATs have limited data speed and hence do not allow real-time voice communication. A voice messaging system like MICROVOX enables transmission of voice message from one VSAT to another.

Hybrid Mail Service. The Hybrid Mail Service facilitating electronic transmission of messages from one post office to another using the VSAT was inaugurated on January 14, 1995. Messages on a floppy after being recorded on a computer will be transmitted to the required post office. The recipient of the message will either receive a print-out or a floppy, and possibly telephonic information about the message if he or she is telephonically accessible. Use of the new service will help in faster and more reliable transmission of messages.

The number of VSAT stations will also be increased by 20. At present, only Delhi, Bombay, Bangalore, Madras, Patna, Lucknow and Shimla are covered by this service.

VSATs can be very useful for video and audio conferencing, broadcast of financial data or news communication of educational value to name a few applications.

VSAT netwroks are going to expand in the coming times, DoT is planning new networks with higher data speeds, for data service as well as voice application. Internationally, newer types of VSATs are being developed. Satellites are planned for low earth orbit (LEO) and middle earth orbit (MEO). While MEO means an altitude of about 10,000 km from earth's surface and typically needs 12 satellites for global coverage, LEO implies an altitude of 50 to 100 km for the orbit. LEO proposals include 'Project 21' of INMARSAT which is expected to be operational by the beginning of the next century. Another LEO proposal is the 'Project Iridium' under which 66 satellites are expected to provide continuous global coverage including mobile service.

VSNL launched the Electronic Data Interchange (EDI) facility in Bombay in 1994, followed by Delhi. EDI is a computer-to-computer transfer of commercial or administrative transactions using a specified standard to structure the data relating to these transactions. Paper work is practically eliminated, as is the need to maintain large inventories at user and supplier ends. It helps in tighter cash management. The Gateway Electronic Data Interchange Service Trade Net set up by VSNL is to be used for DEI service. India is a member of UN/Asia EDIFACT (EDI for Administration, Commerce and Trade) Board whose standards are internationally accepted.

Tifacline is a computer service operated by the DST, TIFAC and CMC. It is linked to databases providing information on processes, products, equipment and policies relating to technology. Video conferencing operates through digital telephone networks. It enables direct face-to-face interaction over long distances. Charts and data can be displayed and discussion can be conducted saving time as well as the trouble of travelling over long distances, and leading to more efficient management control. It could be especially useful in connecting remote areas to central health or educational facilities to access expert advice and consultation.

11
OCEANOGRAPHY

After entering into the nuclear and space clubs India is now poised to become the first developing country to break the big power monopoly in deep sea mining.

India — a peninsula with an extensive coastline and groups of islands — has much to gain from oceanographic research. The new 'Ocean Regime' established by United Nations Convention on the Law of the Sea (UNCLOS) 1982, which has been singed by 159 countries including India assigns much of the world ocean to Exclusive Economic Zones (EEZ) where coastal states have jurisdiction over exploration and exploitation of resources and for other economic purposes. The UNCLOS made declarations regarding (1) the sovereign rights of extraction in the 320 km. Exclusive Economic Zone by coastal states; (2) resources of the deep sea to be governed by International Sea Bed Authority and extraction to be based on the principle of equitable sharing and common heritage of mankind. (To this many developed countries disagreed.) India's coastline is more than 6000 km long and its territory includes 1250 islands. Its EEZ covers an area of 2.02 million sq.km. and the continental shelf extends up to 350 nautical miles from the coast.

Recognising the importance of oceans in economic development and progress of the nation, the Government set up a Department of Ocean Development in July, 1981, for planning and coordinating oceanographic survey, research and development, management of ocean resources, development of manpower and marine technology. The department is entrusted with the responsibility for protection of marine environment on the high seas.

The board objectives of 'ocean development' have been laid down by Parliament in the Ocean Policy Statement of November 1982. The domain of our concern for development of oceanic resources and its environment extends from the coastal lands and islands lapped by brackish water to the wide Indian Ocean. The Ocean regime is to be developed in order to (i) explore and assess living and non-living resources; (ii) harness and manage its resources (materials, energy and biomass) and create additional resources such as mariculture; (iii) cope with and protect its environment (weather, waves and coastal front); (iv) develop human resources (knowledge, skill and expertise); and (v) play our rightful role in marine science and technology in the international arena.

India's research efforts in oceanography has made steady progress. Programmes aimed at developing new resources include: (i) high energy prawn feed; (ii) demonstration farms for edible oysters and green mussles; (iii) hydrocarbon potential of Bengal Fan and Antarctic coast; (iv) cobaltiferrous crust of submerged lettuce; and (v) coastal zone.

Sea food is an excellent source of animal protein. Good management of world fisheries could increase seafood production two or three fold. Phytoplanktons (floating, generally microscopic plants) are the principal plant in the ocean and are at the base of the food chain. The world wide production of organic material by phytoplankton shows four zones within the ocean; 1. Upwelling areas, such as off Peru, California, and West Africa, where production is very high, 2. The shallow water of most continental shelves, which have high production. 3. The Antarctic/Arctic and equatorial water, which generally have moderate productivity because the water is mixed by currents and winds; and 4. The deep-ocean areas far from land which have essentially biologic deserts.

The distribution of zooplankton, which feed on the phytoplankton is similar likewise, locations of the world's major fishing areas follow the pattern of the phytoplankton production of organic material, especially for the near-shore parts of oceans. World's coastal areas contain 50% of the global commercial harvest, the upwelling areas 49% of harvest and other open areas in deep sea produce less than 1% of the global commercial harvest.

The 2.02 million sq km EEZ of India includes roughly 0.2 million sq km under less than 50 metres depth, 0.4 million sq km between 50-200 metres depth over the shelf and hence 1.78 million sq km remained for the offshore fisheries. 50% of the sustainable fishing yield of India's EEZ lies in the inshore zone of less than 50 metres depth, i.e., only in 9% area of the EEZ.

Systematic survey and exploration of marine living resources of the EEZ of India is one of the major objectives. Ocean Research Vessel (ORV) Sagar Kanya and Fisheries and Ocean Research Vessel (FORV) Sagar Sampada are utilised for this purpose. Area of operation of these vessels covers the Arabian Sea, the Bay of Bengal and the Central Indian Ocean down to 20's Latitude.

Sagar Kanya has carried out multi-disciplinary studies in various fields of oceanography in the EEZ. Extensive studies have been conducted on primary productivity and occurrence of zooplankton and phytoplankton, meteorology, air-sea interaction and ocean circulation. Sagar Sampada has carried out extensive studies on pelagic and demersal fishery resources in west and east coasts and also in the waters around Andaman Islands. The survey would help in designing suitable fishing gear, product development for living resources, development of post-harvest technology and preservation techniques, computer mapping of source development and survey of coastal water circulation dynamics.

Sagar Sampada and Sagar Kanya are modern sophisticated research vessels acquired by the Department of Ocean Development.

Brackish water acquaculture is another major resource for food. The coastal areas are being given attention. The programmes for S&T inputs include: (i) semi-intensive prawn farming at Nellore with; (ii) improved extensive shrimp farming in Andaman & Nicobar Islands and (iii) intensive carp culture.

Mineral Resources: The ocean has been repeatedly labelled as the 'last frontier', and claims have been made that metals can be hauled from the sea at 50-70 per cent of the cost of launching; that sea ores are often highly concentrated; and that shelves are rich in petroleum, natural gas, tin, phosphorite, diamons, sulphur and iron, Manganese rich nodules have been halted as a bonanza that would help the economy of developing countries. But still ocean mining has to cope with some problems like availability.

Mineral exploration will undoubtedly be influenced by recent developments in the plate tectonic theory. Three types of environment are areas of major igneous activity; 1. Spreading plate boundaries; 2. subduction zones; and 3. hotspots of rising mantle rock. Ore deposits have been shown to be associated with each of these zones. Mineral deposits of economical interest are quite different from the surrounding sea-floor materials: this allows their spotting and identification based, for example on density, seismic velocity, magnetism, electrical and termal conductivity, indeed polarisation and chemical properties.

Mining beneath the sea-floor includes (a) Petroleum (b) Natural Gas; (c) Coal; (d) Limestone; (e) Tin; (f) Copper; (g) Nickel; (h) iron and (i) Sulphur. Chief prospects for petroleum deposits are youthful margins where some basins contain thick accumulations of sediments. Oil may also occur at convergence zones and near translation faults. Asia remains the continent with greatest promise of discovery.

Manganese nodules (containing mainly cobalt, copper, nickel and manganese") are spherical or acorn shaped ranging between 1 to 20 cm in diameter. They are abundant in areas of low sedimentation rates, such as abyssal plains. Estimates are that as much as 25 per cent of the sea floor is covered by nodules and that 1.5 trillion (1.5 x 10^2) tons are in the Pacific alone, especially in the siliceous ooze area. The nodules of Atlantic and Indian Oceans are generally characterised by having copper, nickel and cobalt below those generally considered economic. The United States has been the prime mover in trying to develop and mine the manganese nodules in the deep sea. Manganese nodules in Indian Ocean cover a large area, over 10 million sq. km. Distribution maps indicate that large areas in the basins east of the Central Indian Ridge contains nodules with a high percentage of manganese, nickel and copper and these appear to be more promising basins. According to recent surveys the Central Indian Ocean Basin's nodules are associated with palae clays.

Considerable interest has been sparked by the discovery of polymetallic sulphides. They occur along the midocean rift valleys, and those found near the Galapagos rift system contain 48% sulphur, 43% iron, 11% copper and smaller quantities of zinc, tin, molybdenum, lead and silver. The sulphides apparently result from submarine hydrothermal precipitation along the northern boundary fault of the rift valley at depths exceeding 2,000 m. It has been suggested that mining the sulphides could be done using the energy of the hot hydrothermal vent waters (360°C). India has located few scattered deposits of polymetallic sulphides in the central Indian Ocean.

The National Institute of Oceanography (NIO), Goa, initiated exploration of polymetallic nodules in the Indian Ocean in 1977. The first sample was collected by the research ship Gaveshant from the Indian Ocean in 1980. Thus India joined a select club of nations — USA, UK, erstwhile USSR, Japan and Germany — who have collected nodules from the seabed. In 1987, India became the first of the four Registered Pioneer Investors to be given exclusive rights to survey, explore and develop a mine site in the deep sea bed of the central Indian Ocean for polymetallic nodules. India is entitled to develop the resources of the deep sea bed in an area of 1,50,000 sq.km.

Efforts have been made to assess the relative concentration and quality characteristics of nodules and broad seabed topography. A multi-beam swath bathy-metric system (hydrosweep) on Sagar Kanya has helped in the survey of the entire pioneer area. High resolution bathymetric maps have been prepared enabling the researchers to identify sea mounts and slopes which cannot be negotiated by mining equipment. The Central Mechanical Engineering Research Institute (CMERI), Durgapur has been given the responsibility of designing and developing a mining system. The technology developed for exploration and extraction of metals will also be useful for bathymetry of the EEZ and deep seabed, exploration and extraction of other seabed resources such as sulphides, metalliferrous muds and phosphorites, and extraction of metals from non-conventional minerals resources such as nickel laterites through hydrometallurgical techniques.

After trying out different process routes for extraction of metals from nodules, three have been chosen for upscaling. The National Metallurgical Laboratory, Jamshedpur has developed a reduction roast Ammonia route and a pilot plant has been commissioned for this. At the Regional Research Laboratory, Bhubaneswar, a pilot plant of 150 kg nodules per day was established on basis of the Ammonia Sulphur dioxide leach route.

The sea water could be the source of an invaluable item-fresh water, the shortage of which is felt in many a region in India. At least the coastal areas can and should benefit from the technology of extracting fresh water from saline.

Several desalination technologies are in operation, but as yet they are not being used on a large scale.

In Avnia village of Gujarat a solar still of 5000 litres per day capacity has been installed. Suitable for small and isolated communities, specially on the coast, the method involves boiling sea water by concentrating solar heat and condensing the steam as fresh water.

Electrodialysis is being used successfully in India. If employs iron-selective membranes for the desalination of brakish water. The energy cost of the process is directly proportional to the salinity, and more economical for salinities below 5000 ppm.

Flash distillation is another method of desalinating sea water. In this method, heated saline water is allowed to flow through a series of chambers which are maintained at different pressures below atmospheric, and progressively decrease towards the end of the series. In each section of the chamber, vapour is released and then condensed over a bundle of tubes cooled by circulating sea-water inside them. Distillate of fresh water produced at each stage is gathered.

Reverse osmosis is the most widely used method. Suitable osmotic membranes are used which reject salts and allow water to pass through when sea water is put under high pressure. Plants of 50,000 to 100,000 litres have been put up in Indian villages. The Central Salt and Marine Chemical Research Institute (CSMCRI) Bhavanagar, Gujarat has done much in this sector, by refining and updating several technologies and passing them to BHEL which is supplying plants of different sizes.

Several scientific expeditions have been sent to the icy continent by India since 1981. The objective of Antarctic research is to identify and initiate programmes of scientific and economic significance besides establishing infrastructure an expertise in the field. In December 1995, India's first commercial Antarctic expedition was launched from Kochi with the aim of catching krill for export to Japan. Sagar Sampada was used for the purpose. The long term programmes include the study of the ice ocean atmosphere of Antarctica in the global environmental context, the Antarctic lithosphere and Gondwana reconstruction, plate tectonics, mineral resources, ecosystems solar terrestrial processes and development of fresh technologies for support system and data bases. Scientific programmes within the above parameters have been carried out in meteorology, geology, oceanography, microbiology, upper atmosphere chemistry, and glaciology. Scientific work is carried out on voyages to and from Antarctica, at the sites of mooring of the ship and on the permanently manned stations. Dakshin Gangotri was the first station to be set up. Mattri, the second one, was established in 1988-89 on a rocky base which stays free of snow almost all year round. The position of Mattri is not suitable for the study of the 'Aurora Borealis'. The NPL, Ahmedabad, has developed an instrument to a study daytime aurorae.

By virtue of its scientific activities in Antarctica, India was admitted as a Consultative Member of the Antarctic Treaty of 1959 in September, 1983. India became a member of the Scientific Committee on Antarctic Research in October, 1984, India also acceded to the Convention on the Conservation of Antarctic. Marine Living Resources from 17 July, 1985. In October 1991 the members of the Antarctic Treaty signed a significant agreement at Madrid. The agreement seeks to ban exploration of oil and other minerals in the Antarctica, known for its fragile eco-system. There is to be protection of Antarctica's delicate flora and fauna for half-a-century. The agreement establishes procedures to assess the environmental impact of all human activity on the continent.

Scientific activities have included geological mapping of the wohlthat massif covering an area of over 600 sq km and collection of samples for petrological, geochemical, geochronological studies; study of various oceanographic parameters and of biomass distribution in different parts of the Arabian Sea, Bay of Bengal and the Central Indian Ocean.

An Antractic Study Centre is being set up by the DoD at Goa. It will formulate national scientific programmes, provide logistic support to expeditions and develop special laboratory facilities among other things.

Marine pollution has increasingly become a cause of concern. Population pressure and rapid industrialisation has led to increased generation of wastes. These wastes reach the sea either directly or indirectly through rivers. This has caused pollution by excessive sewage deposits (organic wastes) and poisonous metals and chemicals (industrial waste) to the marine environment particularly in the coastal waters. Besides these sources, release of agricultural wastes containing pesticide residues and operational releases of ships and tankers containing oil, also cause pollution in the marine environment. Radioactive waste dumped in the sea is also a great danger. All pollutants affect the plants and animals of the seas and upset the ecological balance.

In order to monitor the levels of marine pollutants in a systematic manner as well as to quantify their transport rates from land-based sources to the sea, a well-known multi-institutional programme on Environmental Monitoring and Modelling has been launched by establishing an Apex Centre at Regional Centre, National Institute of Oceanography, Bombay and ten units at expert institutions such as NIO, Goa and its regional centres at Cochin and Waltair, Central Salt and Marine Chemicals Research Institute (Bhavnagar), Centre for Earth Science Studies (Trivandrum). As a supportive measure for the development of suitable methodology on monitoring and modelling, DoD-supported cells at Centre for Mathematical Modelling and Computer Simulation (Bangalore), Regional Research Laboratory (Trivandrum) and National Environmental Engineering Research Institute (Nagpur), have also been established. The programme, after considerable amount of generation of data, is expected to provide knowledge on behaviour of pollutants in the sea and trends of fluctuation of pollutants in the sea which will be useful in the pollution abatement measures.

The DoD has devised a special programme for the coastal zone and islands, with the socio-economic objectives of providing relevant data for managing the reasons and environment of the area. The marine information service is a basic aspect of this programme, and it will provide information on optimal land use design in coastal areas, identify productive uses of certain types of wetlands, provide risk maps of coastal areas for possible rise in sea level and ferocity of cyclones, and identify areas threatened by pollution and erosion so that remedial measures can be taken.

The programme includes a satellite-based real/semi-real time marine information service (MARSIS) under which value-based ocean data products are generated at NRSA and SAC. The NIO validates these data and the Marine Satellite Information Services (MARSIS) centres of DoD disseminates them to the end users. With the completion of Phase I of the project, MARSIS centres have been commissioned at the Institute of Ocean Management, Madras, and Central Marine Fisheries Research Institute, Kochi. Information on potential fishery zone through a network of fax is provided to some of the fishermen cooperative societies in coastal cities and islands. Under COMAPS systematic monitoring and modelling of the marine environment along the entire Indian coastline and islands is undertaken. There is also a programme for precision sea level measurements at selected areas along our coasts and islands using modern tide guages (SELMAM).

Besides continuing with research in the Antarctic, the DoD's achievements include the completion of the survey of the mine site of 1,50,000 sq. km. in central Indian Ocean and the erection of pilot plants for extraction of metals from the nodules and pilot plant compaigns for obtaining material and energy balance. The deep trawi designed and developed at the CIFT has been demonstrated. Forty hot-spots have been identified along the coastline for monitoring pollutant sources. The wave power plant at Vishinjam has been critically analysed. Shore to fishing vessel two-way communication system has been operationalised with the commissioning of the first shore station at Manginapudi near Machilipatnam in Andhra Pradesh. Marine organisms have been identified under the National Project on Drugs from the Sea. A programme has also been initiated to design and develop marine instrumentation and information systems to support resource exploration and harvesting. Acoustic equipment for fish detection, underwater sonar systems, data buoys are some of the instrument systems identified for development.

12

NON-CONVENTIONAL SOURCES OF ENERGY

India has brought the sun down to earth to put cheer into remote villages by providing them water and power. The idea of the strides India has taken in harnessing the solar energy could be guaged from the fact that a variety of systems have been developed ranging from supply of drinking water to providing sophisticated power pack to the first Indian scientific expedition team to Antarctica.

The thrust of the research efforts in the country in the solar energy has been directed both towards solar thermal applications and direct conservation of solar electricity. Utilisation of solar energy is of great importance to India since it lies in a tropical climatic region where sunlight is abundant for a major part of the year.

Until 1979 activity in the field of solar photovoltaic in India was concentrated only on research in solar cells in a few national laboratories. Developmental activity was directed only towards laboratory scale solar cells fabrication, mainly for the space programme. In late 1979, the government initiated a national programme of research development and demonstration of solar photovoltaic energy sources. The area of interest was gradually expanded from solar cells along to solar photovoltaics and from purely scientific research to technological and engineering activities. An integrated approach was evolved for research development, manufacture, deployment, demonstration, education and training with the active participation of a public sector electronics company and various national research laboratories and academic institutions. Whereas scientific research on solar cells and materials was fully supported in the academic institutions, the product development activity was given priority in the public sector company.

In the sixth plan, a significant increase in the allocation and activities has been noticed in the areas of new and renewable sources of energy. A large number of applications of solar thermal conversion both for urban and rural areas have been sponsored by Commission for Additional Sources of Energy (CASE) in this area. Two types of solar collectors have been developed and know-how has been passed on to the industry. A solar energy boosted heat pump suitable for generating refrigeration effect has been developed. Under the demonstration and field trial programme in this area, systems like water heaters and dryers, wood seasoning systems and pumps are being installed. Several methods/models of solar cookers including simple hot box type solar cookers with one reflector, solar oven with multiple reflectors, and solar concentrators have been developed.

The most promising and fast moving solar technology today is that of solar cells, flat metallic blue chips made of highly pure silicon that can convert sunlight into electricity. Though silicon is available in abundance in every country of the world the process of refining the silicon to more than 99.9 per cent purity is extremely complex and its cost is extremely high. This makes solar cells 10 to 15 times more expensive than ordinary ones. The heart of a photovoltaic system is the so-called solar cell which is usually manufactured from a monocrystalline water of high purity silicon. Solar cells have also been made polycrystalline silicon wafers as also amorphous silicon films. The cells are connected in series or parallel combinations to give the desired voltage and current ratings and encapsulated to form modules. The modules are interconnected to form panels and arryas to deliver the required power for any application. The advantages of photovoltaic (PV) technology such as the ease of operation, reliability, minimum maintenance due to lack of moving parts, have been widely recognised. PV systems are thus suitable for installation in rural areas and isolated locations for a variety of applications such as water pumping for micro-irrigation and drinking water supply, community and street lighting, power supplies for microwave repeater station, communications equipment, radio and TV receivers.

Work on photovoltaic technology development in the country began in the fifth plan. Several R&D projects were undertaken at national laboratories and ITTs. A major project on low cost solar cell technology was carried out at the Central Electronics Ltd., which was treated as the lead organisation in the country for PV development. At the start of the sixth plan, the basic technology for a solar cell and module manufacture had been developed and a few systems designed for terrestrial application. Based on these developments a programme comprising of pilot scale production and demonstration and intensive R&D was taken up during the sixth plan.

Researchers in leading laboratories and universities are trying to achieve self-reliance in this vital high technology area which is essentially based on the traditional and mature semiconductor process industry. There are four basic areas of activity in the development of the engineering and technology of solar photovoltaic system which include development of basic raw material, particularly silicon; solar cells and modules balance of system components and complete system engineering.

The Department of Non-Conventional Energy Sources (DNES) has the responsibility for the development, production and application of solar photovoltaic devices. This route of solar energy utilisation is attractive in view of the favourable solar radiation conditions and large requirement for electrical energy for decentralised application. The ease of installation and maintenance lack of noise and pollution, and long life make photovoltaic systems ideally suited for use in remote and isolated areas, forest, hilly and desert regions.

The technology so far developed relates mainly to the processing of single crystalsilicon wafers into solar cells and their design and development for different application. Several improvements and newer techniques in cell and module fabrication have been introduced in production. These include screen printing of front and back contacts, polishing and texturisation, testing and encapsulation leading to reduction the cost of production. Silver paster for screen printing has been developed by CEL. Meanwhile, another public sector undertaking Bharat Heavy Electricals Ltd. (BHEL), has launched a production programme based on in-house technology. Average cell efficiencies ranging between 10 and 11% have been achieved at BHEL and CEL using CZ wafers. Some experimental work using semicrystalline wafers is also in progress. The hand on experience gained should enable the country to acquire and adapt technology and equipment needed to further upscale the production.

In National Physical Laboratory, scientists have developed technology for fabrication of thin cadmium sulphide and coppper sulphide (CdS : CuS) solar cells. These cells are formed over an ordinary window glass substrate in which semiconducting layers in polycrystalline form are built up using vacuum deposition and chemical process. These are totally encapsulated to withstand the climatic changes. To evaluate the reproductibility of the technology and the control over process variables a laboratory level batch production facility has been established. Several hundreds of cells have been fabricated having conversion efficiency in the range of 5-6%. These cells have shown remarkable stability over a number of years.

The NPL has also established facilities for testing the thermal performance of solar collector. Facilities have also been established to test the thermal performance of solar collectors including a comparative testing of a collector with reference to a standard collecting. Besides measuirng experimentally the efficiency of a collector, facilities exist for determining the physical parameters such as the refractive index and absorption coefficient of the glaxing, thermal conductivity of back and edge insulation. Using these and knowing the design parameters, the theoretical efficiency could be determined.

During the sixth plan, the DNES implemented a comprehensive programme of R&D pilot production and demonstration relating to PV technology. The programme led to development of indigenous technology for production, creation of initial infrastructure for the manufacture, installation and maintenance of PV systems and promotes awareness among potential users.

An indigenous plant to produce pure silicon was commissioned at Mettur in Tamil Nadu in 1986. The plant envisages annual production 25 tonnes of polycrystalline silicon and conversion of part of it into single crystal ingots and wafers — the backbone for a host of electronics devices and solar cells that convert sunlight into electricity.

The indigenous process uses high pure silicon tetrachloride (STC) as raw material. It is reacted with hydrogen gas which reduces STC into metallic silicon that gets deposited on a U-shaped heated tungsten filament inside a double shell reactor. the silicon produced at the plant costs between Rs. 750 and Rs. 850 per kilogram which compares with imported silicon. The silicon project is the outcome of a unique 15 year collaboration between Mettur Chemicals and two scientists of the Indian Institute of Science — Prof. C. Suryan, a physical, and Prof. A.R. Vasudeva Murthy, a reputed chemist considered to be the father of silicon industry in India.

An integrated programme on amorphous silicon cells made steady progress. Plasma deposition facilities for experimental work have been established at the Indian Association for Cultivation of Science, Calcutta, Indian Institute of Technology, Delhi, and University of Poona, Silane gas, which is an essential input to amorphous silicon production, has been produced in small quantities at IIT, Kharagpur, and IIS, Bangalore, and is being tested for its quality.

Research on the other aspects of PV technology such as photoelectrochemical solar cells, thin film devices based on compound semiconductor materials such as copper indium diselenide, on implantation and laser annealing systems development continued in several institutions. The DENES is making efforts to promote and evaluate indigenously produced silicon to assess its suitability for PV production. Solar cells and modules made from such materials have been found to be comparable in performance to devices made from imported silicon.

At the Indian Institute of Technology in Kharagpur, researchers are working on ways to change the farm waste into a high-tech energy product. They have found that rice husks contain about 18% silica — commonly known as white ash. The researchers are now working on a cheaper way of reducing silica to fine grain silicon. They hope to use recyclable aluminium in the reduction process and polycrystalline ingots ready for use in solar cells.

There are at present two programmes in operation for production of solar cells and modules in the country. The first of these at the Central Electronics Ltd. and another at BHEL, where a production capacity of 1 MW have been installed.

A five year National Solar Photovoltaic Energy Demonstration (NASPED) Programme was initiated at CEL in October 1980 with an outlay of Rs. 11.97 crore. The programme, which was approved by CASE in May 1981, envisages the establishment of pilot scale production facilities at the level of 1 MW by end of this year, the development of PV systems for a variety of application and the demonstration/development of these systems for purposes of field trials, evaluation and popularisation. The programme sought to upscale the existing production capability and demonstrate the viability of PV energy in Indian conditions.

There are at present two programmes in operation for the production of solar cells and modules in the country. The first of these at Central Electronics Ltd., where a production capacity of 600 KW per year have been installed by the end of March 1984. The actual production under NASPED has been behind the schedule by over six months during the last 2 to 3 years. As against a target of 150 KW for 1983-84, the actual production at CEL is likely to be around 115 KW. A target of 400 KW is being envisaged for 1984-85, which was the figure originally set for the period September 1983 to October 1984. Among the reasons for shortfall in production have been difficulties and delays in developing newer process techniques and applying them for production, shortages of certain raw materials, distribution due to shifting of operations to a new building. Acute shortage of silicon wafers has hampered the production in 1983-84. Two factors have contributed to this shortage, firstly, a shortage of pure silicon in the world market, and secondly, import of silicon wafers required licences under automatic and supplementary licence procedures as per the import policy for 1983-84. Based on the total experience with NASPED so far, vigorous efforts will be required to achieve the targeted production level of 1 MW by September 1988. The second production programme is being implemented at another public sector undertaking, BHEL, with a target of reaching 250 KW by 1988.

About 40 R&D projects covering materials cell fabrication development of systems, concentrating newer types of cells have so far been funded by Department of Science and Technology and DNES. Detailed investigations have been made on the fabrication and evaluation of thin film solar cells including CdS-CuS cells. Work on extraction of silicon from rice husk and metallurgical grade silicon has been carried out with encouraging results. Technology for casting of polycrystalline ingots up to 50mm diameter has been developed and efforts are being made to increase the size. Recently an integrated programme on amorphous silicon technology has been formulated and five projects, including two on saline preparation, have been sanctioned. Several projects in photoelectrochemical solar cells have also been funded by DNES.

There is a significant demonstration/field trial component in the NASPED programme. As part of this, water supplies for offshore platforms, community radio and TV sets, and street lights units have been developed and demonstrated in various parts of the country including Leh. Over 100 solar pumps and 50 other systems have been supplied by CEL. The pumps are being used for microirrigation, drinking water supply as well as training and demonstration purposes.

The first experiments is rural electrification using PV technology have begun recently. Salijipally in Andhra Pradesh became the country's village to be electrified using solar photovoltaic systems. A 7.5 KW power supply system has been in operation since November 1983. Street lighting and other units are proposed to be installed in some tribal villages of Tamil Nadu soon. The feasibility of installing such systems in Jammu & Kashmir and Karnataka is being explored.

According to an official survey, the country has an excellent ocean thermal energy conversion (OTEC) potential and some of the best sites in the world are situated of the Indian mainland and near the islands Lakshadweep, Andaman and Nicobar. The total OTEC potential round India is estimated to be more than 50,000 MW which is about 153% of the present total installed power generation. A project is being launched by IIT, Madras, in collaboration with other national organisations.

There are other renewable sources such as hydro power, ocean energy, geothermal energy in all of which the country has programmes to succeed and for renewable sources of energy to make a significant impact in the country, it will be necessary not only to mount large scale programme sponsored by the government but also to involve the industry and voluntary agencies and other organisations in this task.

Funding by the central government for installation/use of solara thermal systems or device including community solar cookers varies from cent percent in case of its institution to 33.3 per cent in case of family type solar cookers. Solar photovoltaic water pumping systems and community lighting systems are being supplied at a subsidised price of Rs. 25,000 to state governments and other agencies. Some state governments and union territories are also giving subsidies on various solar devices.

In a bid to provide a simple and cheap method of energy production, Indian scientists have developed a number of designs of biogas plants which make use of wide range of agricultural wastes like animal dung, human excreta, vegetable waste, water hyacinth and produce fuel in the form of gas, simultaneously with high quality manure.

About a dozen of designs, incorporating distinctive features either in the digester or gas holder have been developed in the country by various research institutes.

The research and development efforts in biogas have yielded useful results. Development of different designs of fixed dome biogas plant, development of ferrocement digesters, successful introduction to FRP to biogas technology and replacement of steel holders, rationalisation of biogas plant designs based on different agro-climataic conditions such as ambient temperature, resulting in reduced capital cost for construction of the plant, development of cheaper biogas plant designs which use lesser quantity of cement and masonry are some of the major achievements which deserves special mention.

Besides, the scientists have developed biogas plants which can operate on a variety of feed materials such as night soil, water hyacinth, agricultural wastes including banana stem, deoiled castor cakes, willow dust and food waste. Development of specialised and improved micro-biological cultures for biogas generation from deoiled castor cakes and other non-edible waste and residues, development of engines which can run on biogas alone (100 per cent biogas) and dual-fuel (20 per cent) diesel: 80 per cent biogas and successful integration of solar heating systems with biogas plants for improving gas generation during winter are among the other achievements.

There are four types of biogas digester designs; floating gas holder digester, fixed-dome digester, flexible bag digester, and experimental small and cheaper digesters, are in vogue today.

The floating gas-holder digester was the most popular digester design in the country in the early years of its introduction. Problems of corrosion and gas leakage and the high construction cost of the metallic gas-holder caused the collapse of its popularity. It is expected that the development of gas-holder made of a cheap corrosion-free durable material may help regain the popularity of this design. For this reason the development of a cheap and good quality floating gas holder made of fibreglass may be further investigated.

The fixed-dome digester is the most popular design in India today. Its popularity is mainly due to its lower construction cost, local availability of the required building materials, its hidden underground structure, and the non-existence of the expensive floating metal gas-holder with consequent reduction in maintenance costs. On the other hand, a plastered masonry dome is bound to develop cracks leading to gas leakage within a few years; and repair is a difficult job. To make this fixed-dome digester more effective, and scientists say the search for a cheap and appropriate lining to seal the inside of the dome should be intensified.

Very little data are available on the experience with the plastic bag digester in the country. However, flexible bag digesters appear to be very promising. Few operational problems, easy and cheap installation, high durability and stable gas pressure are its major advantage.

Smaller and cheaper digesters are expected to be more appropriate for the poorer groups. Moreover, the demand for smaller digester feed in these digester corresponds better with the mostly smaller quantity of cattle dung available with the villages. Portable digesters are also under development to overcome the immovable character of conventional digesters.

Experiments conducted at the National Environmental Engineering Research Institute have shown that cattle dung and water hyacinth, mixed in the ratios of 3:1 and 1:1 at an organic loading of 1.6 kg of volatile solids per cubic meters per day and a detention time of 30 days, can yield 30% more gas than with cattle dung alone.

It is a horizontal plant with a steel floating dome a vertical fed pipe at one end for easy feeding of the vegetable matter by pushing with a bamboo; a sloping roof and floor which helps the passage of light weight vegetable matter and its sludge towards the outlet; and a big outlet of suitable dimensions for easy disposal of partly decomposed sludge.

The Regional Research Laboratory, Jorhat, is now engaged in standardization so as to develop field-scale prototype digesters and optimum operation design. Two demonstration plants of 300 litres capacity having different dimensions and feed system were run for several months in the Central School premises of CMERI colony and in the institute's campus at Durgapur.

A system of using ferrocement precast segmental units for constructing digesters for biogas plants (3 to 8 meter cubic gas per day) has been developed and tested at the Structural Engineering Research Centre (SERC), Roorkee. The digesters are assembled with 4, 6 and 8 precast segments 15 to 18mm thick. The wiremesh and wire reinforcement projecting on sides are used for connecting the segments when placed in a vertical position. The mesh laps from the joint ribs and also accommodate the inlet and outlet pipe junction with the digester wall. A rich cement-sand mortar is applied over the mesh joint areas and ribs are formed. Extra reinforcement could be provided in the joints and radial bands, projecting out at top, middle and bottom positions. The ferrocement (cement mortar reinforced with steel wiremesh layers) gas holders, designed for biogas plants of 2.6 meter steel capacity gas production per day have also been developed in the country. The holders are 40 to 50 per cent cheaper than steel gas-holder.

The National Environmental Engineering Research Institute (NEERI), Nagpur, has designed a night soil digester to prevent indiscriminate dumping of night soil and its unhygienic handling on farms in raw condition. The digester oxidizes organic matter rids the material of obnoxious odour, eliminates pathogenic organisms, provides sluge as manure, and yields methane gas for domestic use. The technology has been adopted at six places; Ramakrishna Mission, Coimbatore; Agriculture College, Rahuri, Maharashtra; Sulabh International, Patna; Burujwada Village near Nagpur; Garden Reach Municipality, Calcutta; and Delhi Contonment Board, Delhi. While in the Patna plant only night soil is used in others, mixtures of night soil and cattle dung are used.

An electrochemical method of preventing corrosion has been developed by the Central Salt & Marine Chemicals Research Institute, Bhavnagar, to reverse the corrosion process. The method is cheap and can be easily adopted by farmers in villages. It requires no supervision or maintenance.

The technology for production of biogas from cowdung and other animal wastes has been known in India for several decades now. The advantages of biogas as a fuel in the kitchen include elimination of smoke, improved in the sanitary and environmental conditions of homes and villages production of manure after digestion of wastes. Under a national project for biogas development implemented by DNES over 0.65 million plants were set up till December 1986. This would save over 2.5 million tonnes of wood equivalent Rs. 1,000 million a year. This will also produce as residue, excellent fertiliser valued at another Rs. 1,000 million a year.

Chulhas (woodstoves) that do not emit smoke and use less wood are fast replacing traditional cookstoves in rural households ushering what is described as 'smokeless revolution' in India's country-side. About one million smokeless chulhas have so far beeen installed and 6,000 villages have become smokeless.

About 40 per cent of all the energy consumed in India goes for cooking food. The traditional woodstoves, made of mud and brick, in India's 112 million homes annually burn 133 millions tonnes of firewood and over 200 million tonnes of cowdung cake. But the stoves are so inefficient that rural women spend long hours in collecting firewood and working in kitchen filled with smoke due to unburned particles that Indian scientists say are carcinogenic. The improved cookstoves, having a combustion efficiency of more than 15 per cent have drastically cut down consumption of firewood. And because these stoves are designd to burn the fuel completely the kitchens are smoke free.

Some 51 models of improved cookstoves have been developed and introduced by various social organisations and research institutions like the Indian Institute of Technology (IIT) in Delhi, the Central Power Research Institute and the Indian Institute of Science (IIS) in Bangalore. The cookstoves vary according to the regions, type of food cooked, the fuel used. But the designs incorporate optimum size for combustion chambers, air inlets, grates for improving air circulation on fuel bed, dampers for draft control and baffles to improve heat transfer. Some models for permanent fixing in kitchens with or without chimneys. Others are portable type and are made of metal sheets or cast iron or a combination of the two. Apart from fuelwood most of the models can use cowdung, coal or pelletised agro-waste. Thee three-pan 'Astra Ole' stove designed by IIS requires only 60 grams of fuelwood for cooking one kilogram of rice in contrast to 400 grams consumed by traditional stoves. All new designs are tested and certified by the testing and training centres specially set up at IIT, CPRI and the Punjab University of Agriculture in Chandigarh.

This revolution has caught up in all of India's 25 states and union territories. The desert state of Rajasthan boasts of 135,000 improved stoves and 565 smokeless villages followed by the northern state of Haryana where 84,000 of these cookstoves installed in just one year have vastly changed the lifestyle of women in 200 villages. Funded and subsidised by DNES, the project is implemented by voluntary organisations, state nodal agencies and women's organisations such as the All India Women's Conference. DNES has so far conducted 1,547 training programmes to train people in the construction of the stoves. The master trainers in turn train men and women in their villages in 1986 alone some 81,000 persons had received training and DNES expects to have a workforce of 100,000 workers by the end of 1987. Each worker turns out five chulhas a day and earns Rs. 25 in wages.

While the fixed type chulhas are installed by government agencies and voluntary organisations, portable models are marketed at prices ranging from Rs. 38 to Rs. 63 by 12 private companies that have sprung up to cash in on the smokeless revolution. An incentive to buyers, the government has abolished excise duty on fuel efficient wood burning stoves.

DNES has also developed community chulhas capable of cooking for up to 200 persons with fuel efficiency as high as 35 per cent. These have been recommended for hostels, hotels and canteens. The Tamil Nadu Government which has a mid-day meal programme for school children is planning to instal the improved community chulhas in the 33,000 schools whose mid-day meal project is being threatened by fuelwood shortage.

It is hoped that cookstoves, like bullock carts, will dominate the Indian rural scene for many more years. The new stoves are one important means of stretching the fuelwood resources and removing a major drudgery of rural women.

Biomass, in its variety of forms, is a potential source of renewable energy for use as solid liquid and gaseous fuel. It offers an important alternative to provide substitute to petroleum products. A number of activities relating to assessment, production, improvement, conversion, and utilisation of biomass were undertaken during the last four decades.

Increasing productivity of fuelwood trees and shrub species grown on problem soils - saline, alkali soils, degraded hill slopes, coastal alluvium, sand-dunes, laterite and lateritic soils, black soils, ravine lands - is an important national problem. For this purpose, it has been planned to set up a chain of Biomass Research Centres under different agro-climatic conditions of the country. In addition to the three centres already sponsored at Garhwal University, Srinagar, NBRJI, Lucknow, and MIKID, Madurai. Two more such units at M.L. Sukhadia University, Udaipur, for Western Region and Purulia, West Bengal for eastern region have been approved.

The Biomass Research Centre at Garhwal University, Srinagar, is engaged in screeing the mountainous tree species for the biomass potential. The studies include germinability of seeds, growth behaviour, light harvesting potential, photosynthesis and wood characteristics. Nearly 70 species have been collected and investigated for optimum temperature requirement under tropical and temperature conditions.

In most of the leguminous species, a high percentage of germination is recorded. However, Albizzia lebbeck, a promising fuel and fodder species, exhibits dormancy in fresh seeds and requires hot water treatment for a early and uniform germination. The seeds of Toona ciliata, Cedrela, Alnus indica, Holoptelea integrifolia, Alnus nepalensis and Morning oteifera gradually lost their viability. Observations on growth behaviour indicate that Prunus cerasoides, A. Lebbek, Terminalia bellerica Dalbergia sissoo, Sapindus muscorosi, O lea glandulifera and Toona ciliata, are promising fuel and fodder species for high biomass production in hilly areas. A mathematical equation based on the optical properties of leaves, by which the biomass production potential of a species under a particular climate can be predicted has been derived. The centre has designed styrofoam seedling tray measuring 30 cm in length and width and 10 cm in height. It accommodates sixteen seedlings and weighs only 5 kg with fully grown 6 months old seedlings. The tray can be used for raising as well as transportation of the seedlings. Its use minimises the overall investment per seedling and increase the chances of survival of the plants due to undisturbed root system before planting.

The Biomass Research Centre at NBRI, Lucknow, was set up to carry out studies on comparative evaluation of fast growing tree species suitable for saline-alkaline soils. In a high density trial Terminatia arjuna seedlings planted at 0.500.66 (corresponding to a density of 30,000 plants/ha) were significantly taller at an age of 3 years in comparison to widely spaced plants (10,000 plants/ha). The stem diameter increased with increasing density but differences between treatments were not significant. Soil up to 30 cm depth in the rooting zone, irrespective of density of planting showed a decrease in pH and exchangeable magnesium, an increase in organic carbon, exchangeable sodium, potassium and calcium content. Electrical conductivity did not indicate any significant variations.

In another experiment with firewood shrub species, Sesbania aegyptiaca proved significantly superior in terms of total biomass. Biomass yield tended to increase with an increase in plant population. Soil pH decreased and organic carbon content increased irrespective of the species. A pronounced increase in Ca/Na ratio was recorded under Sesbania aegyptiaca with highest value at a density of 75,000 plants/ha. Biomass production per unit area was highest in Prosopis juliflora. Amongst the five successful species with survival above 86 per cent, the most uniform average heights were achieved in Acacia nilotica.

The Biomass Research Centre at Madurai Kamaraj University, Madurai, is designed to carry out investigations in three major area of research selection of suitable fuelwood species for the semi-arid region; development of nursery technology and mass production of seedling; and selection of suitable rhizobial strains. Out of nine species selected, Leucaena was found to be more aptly suited for plantation under semi-arid condition on degraded land in Madurai district. It was observed to be significantly drought resistant. Photosynthesis and a number of related physiological parameters were correlated with the growth and bioproductivity of the selected woody biomass species. The species with high chlorophyll content showed a high degree of correlation in their photosynthetic rate. The use of the stem circumference measurements enabled to follow accurately the annual rhythm of the trees and investigate their wood productivity. A new phenotype of Leucaena was selected detailed investigations which is four times more productive than the normal one. As a part of the demonstration activities of the Centre, seeds and seedlings of selected fuelwood species were supplied to four taluks of Madurai district for Rural Energy Plantation Programme for power generation.

A Biomass Research Project for Coromandal Coast was sponsored at the Institute for Offshore and Coastal Research of Visakhapatnam, Andhra Pradesh. The farmland at Lankelapalem with an area of 45 ha possesses four types of soils, viz., sandy loam, loam, clay and gravel. Seven species of legumes viz., Albizzia lebbek, Cassia siamea, Leucaena leucocephala, Pithecellobium dulce, Prosopis juliflora, Sesbania grandiflora and Acacia and three non-legume (Ceiba pentandra, Casuarina equisetifolia and Eucalyptus hybrid) were planted at the farm. In studies on comparative evaluation of 9 species Sesbania grandiflora appeared to be the most promising reaching a height of 8 meters within a period of 16 months only. All the transplants were found to be performing better in the loamy soil. The survival rate was observable around 85% at all the levels.

In order to correlate Yield and development of plants with environmental changes, a Data Collection Platform (DCP) has been established with the support of the Indian Meteorological Department at the Research Farm. This platform is equipped with instruments for measuring various climatic parameters on a regular basis. The data collection and retrieval are monitored by geosynchronous satellite INSAT-1B randomly in a nurst mode every hour.

Under the project, "An Investigation on the Fuelwood Yielding Plants with Special Reference to their Improvement Through Cytogenetical and Culturing Techniques", studies were carried out on the following aspects; assessment of chromosome status of the selected fast-growing fuelwood plants ; genetic, improvement of the desirable plants through polyploidization and induction of mutation by application of different physical and chemical mutagens; rapid clonal propagation of the selected plants through in vitro culture technique; exploration of the possibility of improved seed germination technique for faster multiplication; and the study of the growth rate and biomass yield potential of the selected fuelwood plants in different agroclimatic zones.

Detailed cytological analysis of the four species has revealed that Leucaena leucocephala is a polyploid species having a very high chromosome number (2n = 104). Sesbania gradiflora has the lowest number of chromosomes (2n = 24) whereas Acacia auriculiformis and Prosopis juliflora have 2n = 26 and 52 respectively. In the species chromosomes are minute in size. Difficulty in seed germination as well as profuse cytoplasmic inclusions poses serious limitations in working out of karyotypes in other species. Complete regeneration in vitro condition has been successfully done in case of Sesbania grandiflora. Profuse shoot multiplication in Leucaena leucocephala should be utilised for rapid propagation after successful rooting becomes possible. A large number of plants with vigorous growth have been recorded after gamma-irradiation in Leucaena leucocephala, Acacia nilotica and Prosopis Juliflora. This opens up the possibility of higher biomass yield and faster growth after irradiation in other materials also. Similar phenomenon has also been noted after treatment with certain chemical mutagens. Final screening of the mutants may yield some useful variants with increased biomass yield. Induction of polyploidy has been successfully done in case of Sesbania grandiflora. Assessment of biomass yield and growth behaviour revealed that Acacia auriculiformis, Sesbania grandiflora, Leucaena leucocephala and Dalbergia sissoo have maximum potentiality in all the regions. Rapid seed germination technique has been standardized in some of the materials.

A number of energy-rich leguminous trees and latex producing plants are under investigation at the Applied Botany Section, Kharagpur. Calotropis gigantea (Akanda), a hardy weed of arid and semi-arid regions, produces latex (M.W. 10,000) which is a potential source for hydrocarbons. Methods have been developed for micropropagation and regeneration of plantlets from in vitro callus cultures of Albizia (Siris), Samanea (Rain tree) and Peltophorus (Copper pod). Experimentally it has been shown that for large scale micropropagation, as many as 30 to 40 plantlets can be regenerated from a stem segment of 0.5-1.0 cm during a period of six weeks only. The usual high mortality of the plantless on transfer to soil can be overcome by induction of extensive rooting and hardening under shade. It has been established that non-committant caulogeny and rhyzogeny is the chief reason for high mortality of potted plants.

Jojoba (Simmondsia chinensia Link; Schnd), a supplementary source of lubricating oil, has been tried in the semi-arid and sub-humid climates in the western and eastern regions of the country, respectively. The performance of the species is significant with two fruiting cycles and oil contents maintained at about 50%. Thirty-six selections in the west coast and 10 selections in the east coast have been made with clear distinct plant types. Multi-locational trials have been initiated with the help of government departments, universities, research institutes and private entrepreneurs. In one of the plantations in Gujarat, flowering and fruiting has commenced in two-year-old plants on a private farm.

Morphological and biochemical work has been initiated to identify the sex at an early stage so that sex differentiation should not be constraint in making large scale plantations. Vegetative propagation by air-layering, stem cutting and tissue culture have been undertaken. The progeny P, P and P have been planted and the observations of mother plant of the 36 selections made in the semi-arid climate.

Indian coastline stretches about 6,000 kms and sea-weeds are utilized for agar and alginate extraction. The utilization of the seaweed biomass as a substrate for biogas has been sponsored at Marine Alga Discipline, Central Salt & Marine Chemicals Research Institute (CS & MCRI), Bhavnagar. Experiments carried out on the biogas from sea-weeds indicated that Sargassum tenerrium, a species of brown sea-weed alongwith 10 to 25% cowdung and marine microbial culture can yield 90-100 litres of biogas per kg (dry sea-weed) over a period of 90 days only.

Studies were carried out on hydro energy of biocrude samples obtained from Euphorbia nerifolia and E. ryleana and Cryptostegia grandiflora and Calotropis gigantea using Silica alumina based catalyst ratio of 10:1 at IIP, Dehradun. The operating temperatures and pressure were 380-450°C and 80-150 cm, respectively. In a typical run, products obtained from the latex processed biocrude were : gases (6.61) naphtha (23.85) kerosene (21.97) gas oil (18.22) heavies (26.47) water (0.95) coke and losses (1.93%). In the case of biomass processed biocrude the products were gases (6.46) naphtha (16.17) kerosene (28.26) gas oil (16.17) heavier (12.93) water (90.97) and coke and losses (19.04%). Over 90% of the hydrocarbons in the lower boiling products have been identified. These studies clearly demonstrated the feasibility of obtaining petroleum hydrocarbons from renewable sources.

Development of products using jojoba oil or its derivatives was continued at IIP, Dehradun. Using neat jojoba oil as a base stock for two stroke engine oil formulation was developed. The properties and screening test of this formulation were found to be comparable to those of commercial two stroke engine oil with respect to wear, scuffing and deposits. Conditions for the preparation of sulphurised jojoba oil were standardised and the product was evaluated for wear and load carrying capacity. Comparatively study with commercial additives was also carried out. The sulphurised jojoba oil so developed was found to have promising application as an extreme pressure (EP) additive, a formulation of industrial gear oil developed and the properties of this oil were evaluated. The formulation meets all the requirements of IS-8406 and it has been found to be comparable to commercial oil. Jojoba oil derivatives with higher sulphur content for use as EP additives are under evaluation.

Vegetable oils can serve as possible substitute or supplement to diesel oil. Vegetable oils have calorific value and ignition quality approximating to those of diesel oil; however, their viscosity and carbon residue is very high. This leads to pumping and vaporisation difficulties in engines and also heavy smoke emission in the exhaust. In order to overcome these problems, two different approaches, namely making the diesel engine more adiabatic and esterification of the vegetable oil with methyl or ethyl alcohol have been tried at IIT, Madras. In the first case, namely making the engine like piston, cylinder liner and cylinder head are thermally insulated either by ceramic coating or by air gap. With the heat losses in engine drastically reduced, the air temperature after compression increase substantially over that of the normal engine. Hence, the evaporation and ignition of vegetable oils is made easier. In the second approach, vegetable oil is reacted with ethyl or methyl alcohol in the presence of catalysts to form the eather of that oil and glycerine. By this process, the viscosity is brought down to the level of diesel oil and also the carbon residue is reduced. The use of karanji oil, salseed oil and rice bran oil is under investigation at IIT, Madras. An engine has been converted to be adiabatic and Karanji oil has been tried on this engine. The ignition delays the rate of pressure rise and the peak pressure are higher than for the normal engine, indicating the benefits of preventing heat losses.

Three projects on the development of a village level gasifier based on wood and agricultural residues for irrigation purposes have been undertaken at IIT, Delhi, and Jyoti Solar Energy Institute, Gujarat.

A gasifier-cum-clean-up system has been developed at the Chemical Engineering Department, IIT, Delhi. The gasifier unit is very simple in construction which will enable even an unskilled farm labourer to operate and maintain it easily. A wetted wall cyclone has been designed for effectively cooling and cleaning the producer gas from the gasifier. A simple batch drum charring unit and also a semi batch charring kiln have been designed and developed for charring biomass of various types. These units are under successful operation. The gasifier system has lodged more than 200 hours of trouble free operation, producing a gas of 820 KCal/m^3. A 5 HP C.I. engine pump system supplied by Project B team from Mechanical Engineering Department, IIT, Delhi, has been successfully operated at about 70% diesel replacement.

Research and development work on efficient utilisation of producer gas in combustion engines has been in progress at the Mechanical Engineering Department of the Indian Institute of Technology, Delhi. Existing 3-5 HP petrol engines and diesel engines have been modified to operate on producer gas fuel. Extensive performance and emission tests have been carried out on 5-20 HP stationary and automotive type of petrol engines duly modified for 100 per cent producer gas substitution. The 5 horse power diesel engines of the type widely used for irrigation purposes have been modified to operate with up to 80 per cent full load energy substitution by producer gas. Engine design and operating parameters such as injection timing compression ratio, intake system, etc. have been experimentally optimized to achieve efficient combustion of producer gas with an annual saving of more than Rs. 5,000 in diesel cost per engine pumpset/gen set of 5-6 KW rating on a 10 hour per day operating basis.

Under the gasifier programme, Jyoti Solar Energy Institute, Vallabh Vidyanagar, has developed a gasifier system capable of running 3.7 KW engine-pumpset for irrigation and installed a demonstration unit in a farmer's field at village Johr, 7 km from the Institute. It is a down draft gasifier with facility for air pre-heating. The system has run for about 500 hours without any technical problem. It consumes wood at the rate of 3.4 kg/h and gives an average diesel replacement of 60 per cent. The operating cost of the unit is 70P per KWh of shaft power whereas with electricity and diesel, the corresponding values are Re. 0.87 and Rs. 1.07.

A small scale magneto hydrodynamics power generator set up at Tiruchirapalli in Tamil Nadu is providing data for design of bigger MHD plants that will produce power cheaply and operate of a greater efficiency than existing coal or nuclear plants.

The experimental run for power generation was carried out by the Bharat Heavy Electricals Ltd. in collaboration with the Bhabha Atomic Research Centre, Bombay.

The principal of MHD power generation involves direct conversion of thermal energy into electrical energy and the process is being experimented upon several countries like USSR, USA, Japan, China, the Netherlands, Australia, Poland and Finland.

According to a senior BHEL scientist the science of converting thermal energy to electricity through MHD involves the expansion of superhot (2,800 K), electrically conducting gas against the retarding force of a strong magnetic field to produce electric power directly. In effect, the turbine and generator are combined in MHD into a single unit but without any moving parts.

It is well established that flame temperature of fuel oil, coal, natural gas or power gas in reaction with preheated air at 1, 200°C, and 5 atm. is about 2,500°C. Thermal ionization occurs at this temperature, but the level of electron density is insufficient to induce economically viable electric conductivities in the combustion products. However, the addition of an easily ionized metal seed, such as cesium or potassium, in a concentration of about 1% during the combustion process provides the required conductivity in the range of 10-100 mho/m^3. The apparatus which accomplishes these arduous tasks is the MHD combustor.

The combustor is followed by a nozzle which accelerates the plasma to subsonic or supersonic velocities. Deceleration and electrical energy extraction occurs in the MHD generator, which is positioned in a high magnetic field of 4-5 tesla. The cooled exhaust-still at high temperature (2,200K) carries the remaining (a large) portion of the input energy and this is used for further power generation in a rankine plant downstream. After passing through the stream generator, the exhaust is scrubbed to separate the expensive seed and is then ejected via the stack. The seed is collected and processed to remove impurities and then recycled. Later DC electrical output is inverted to AC and heat losses through the combustor and MHD generator are recovered in separate loops.

The coal based MHD researchg project in India, sponsored by the Department of Non-Conventional Energy Sources, aims at the creation of a suitable base for research and development work in the field of MHD power generation by setting up a thermal power level of 5 MW.

BARC has developed most of the key components of the facility including the MHD generator, a 300 tonne of electromagnet, special plasma diagnostic equipment, high-temperature ceramic materials and as control and instrumentation system for the experimental plant.

The plasma generation experiments, carried out during March 1985, yielded MHD-grade plasmas at 2,700K and very high electrical conductivity. Following the plasma generation run, the first successful power run of the project was accomplished in August 1986. A cold oxidiser with 60% oxygen enrichment was used during the power run.

The experiments conducted so far have shown that the new technology could increase the energy conversion efficiency from the currently obtained 29-30% in India to 46%. Further technological advances may lead to more attractive efficiency. Besides, an annual saving of 8,00,000 tonnes of coal per 100 MW of installed capacity can be achieved at a capacity utilisation of 70%.

Studies further found that the quantity of heat discharged from an MHD-steam plant is less than 50% of that from the best conventional plant and less than 25% of that from a nuclear power plant. Hence, MHD-steam plant could be located where fuel is available in plenty but cooling water is limited.

The MHD-project initiates a two-year experimental programme. These experiments are expected to generate technical operational and economic parameters in the field of MHD power generation which will be useful in further upscaling and eventual commercialisation of this technology.

The experts believe that MHD could become a major electric power generation technology; in fact, the commercialisation of this process is anticipated as early as middle of 1990s. The aim and effort in MHD development that has so far received world-wide attention, and has the most immediate large-scale application, is focused on the prospect on improved energy conversion in fossil fuel based systems to produce power for a utility grid.

Moreover, the costs of conventional fossil fuel fired plant has been rising steadily since the early 1970s because of increased regulatory restrictions on releasing pollutions to the atmosphere. MHD, the experts say, is an enviromentally clean system. Besides, it is more efficient and less polluting energy alternative holding promise of generating electricity at a reduced cost.

In the field Wind Energy Research & Development efforts were oriented towards improvement of existing designs of wind pump system, development of new windmill design and on the development of small battery chargers and wind electric generators.

A design review of the 12-PU-500 windmill was carried out in order to assess the adequacy of the present design for the intended duty. This has led to the identification of design weaknesses and the improvements needed in various sub-systems such as pump, transmission and furling mechanism for better and more reliable performance. A new design of 12-PU-500 windmill has been developed in Gujarat Extensive field trial of the new design is proposed to be taken up.

Under a project sponsored at National Aeronautical Laboratory (NAL) for development and field testing of sail type windmills, different configurations were tried successively for a "trapezoidal" sail type rotor having "self activated twin spoiler" regulation device which had earlier shown significant promise in wind tunnel tests. From the experience gained with the sail type prototypes, the sails were replaced by a 3 bladed metal rotor of 7.5 m diameter, designed for tip-speed ratio of 4:0. The rotor is coupled with 2 HP centrifugal pump. The starting speed is 12-13 kmph and rated speed is reached in 21-22 kmph, when the pump delivers 21,000 lph. The output is particularly high in wind speeds upwards of 16 kmph, indicating that the windmill is more suitable in windy areas. The windmill has been in operation since end of August 1985; initial tests have shown significant regulation capability. Four prototypes of the windmill with minor modifications in the rotor and gear boxes will shortly be installed for field testing.

13

RURAL TECHNOLOGY AND ENVIRONMENT MANAGEMENT

'Rural Technology' is characterised essentially by its inherent capabilities of creating gainful employment, increasing productivity, recycling wastes and byproducts and producing value-added items, thereby augmenting productive work and enhancing income in rural milieu. Human welfare through better housing and sanitation, community services, elimination of drudgery especially among the womenfolk, promotion of decentralised techno-economic systems, particularly for the remote areas and such other aspects of human development are also catered to by the technologies generally grouped under 'Rural Technology'.

The National Research Development Corporation (NRDC), Council of Scientific and Industrial Research (CSIR), Indian Council of Agricultural Research (ICAR), Department of Science and Technology (DST), Khadi Village Industries Corporation (KVIC) and the Department of Rural Reconstruction have done considerable work on the generation, development and demonstration of new technologies in the field of food, water, sanitation, environment, human settlement, roads and communication, energy and natural resources. The area, in which a number of appropriate technologies have been developed are : leather industry, pottery making, processing of cereals and pulses, production of matches, fireworks, soap making hand-made paper, bee-keeping, coir and other fibre making industries, biogas, lime manufacture, blacksmithy and carpentry.

Traditional method to filter water through beds of sand, gravel and charcoal to remove suspended impurities followed by boiling to kill harmful bacteria is not only time consuming but costly too. The method has been simplified by Regional Research Laboratories.

The Indian scientists have developed low cost but equally effective alternative solutions to the capital intensive drinking water technologies. They have developed a number of techniques to make water potable and free from disease germs like bacteria, viruses, protozoa and worms besides being physically clean and acceptable in taste.

The National Environmental Engineering Research Institute (NEERI), Nagpur, has carried out a critical evaluation of selected rural water supply schemes in eleven states and has worked out project proposals for rural water supply covering planning, design, construction, operation and maintenance.

In rural areas where ground-water available is of good quality and at reasonable depth, it is drawn by means of handpump, tube-wells of appropriate design. As this water stands filtered by nature, normally no treatment is given except chlorination for safety. A compact water-treatment plant suitable for rural conditions has been designed and developed by the NEERI. The plant has three concentric cylindrical compartments, the innermost flocculator is equipped with paddles. The alum solution is fed by a constant-head gravity device. The flocculated and settled water is filtered. A bleaching-powder solution is added to the flocculated water before it flows into the sedimentation basin. As a result, there is adequate contact time for disinfection, and the sedimentation basin and filter area, free from organic fouling.

The plant was operated continuously for a year and data were collected using Kanhan river water with a turbidity range of 20.1, 900 NTU. The alum dose ranged from seven to 200 mg/litre. The physiochemical characteristic of treated water conformed to the standards for drinking water. The skill available in the villages was adequate to operate and maintain the plant. The power consumption was nearly one unit per cubic metre of treated water and the estimated cost of treatment per cubic metre was about Re. 1.00.

The Structural Engineering Research Centre (SERC), Ghaziabad, has developed economical and light weight ferrocement water tanks of capacities ranging from 270 to 20,000 litres to provide drinking water to rural areas. These tanks can also be used for storing water required for sanitation. Ferrocement, a highly versatile form of reinforced mortar, consists of closely spaced layers of wiremesh reinforcement impregnated with a rich cement-sand mortar. It is possible to cast ferrocement elements with a thickness of as small as 1 cm. It possesses high resistance to cracking and unlike structures also has high corrosion resistance.

Water tanks developed at SERC are cylindrical in shape and are assembled with prefabricated components. Tanks walls can be produced by using either of the two casting methods developed at the institute. The methods are: semi-mechanised process for producing cylindrical units ; and casting segmental shell elements and assembling these into cylindrical units. While the first process is suitable for producing tanks of up to 2,500 litres capacity, the later process is suitable for tanks to be used in hilly areas or villages where transportation is difficult and also for tanks of larger capacity than 2,500 litres.

The charcoal water filter, developed by NEERI, is an effective technique for the removal of solids, suspended materials and other harmful bacteria. It can filter sufficient drinking water for one family.

The Central Glass and Ceramic Research Institute (CGCRI), Calcutta, has developed water filter candles, which can be fitted to domestic water containers, including earthern pitchers. The candles remove harmful bacteria from the raw water. The Regional Research Laboratory (RRL), Jorhat, has developed a process for the manufacture of domestic water filter candles that can be fitted to domestic water containers, including earthern pots and plastic buckets. Suitable grade of clay, a binder clay and bacteria inhibitor are grounded separately and passed through a sieve. These substances are then blended and mixed with aqueous solution to obtain slurry. Candles are then prepared from slurry in moulds, dried and baked. This technology has been licensed to 16 parties and some of them have gone into production.

A chlorination pot for wells, which effectively chlorinates water in open dug wells up to three weeks, has been developed by scientists of NEERI. A pot of seven to eight litres capacity is taken and 6-8 holes of half a centimeter diameter are made at the bottom. The holes are covered with stones or pebbles of 2.4 cm size. This layer is covered with pea gravel of smaller size. This layer is covered of bleaching powder and three kg of coarse sand is prepared and spread over the pea gravel. Finally, the pot is fitted with pebbles or stones up to the neck. Now the pot is lowered in a well with the help of a rope approximately one metre below water level. The pot is enough for a community well where 900-1,300 litres are drawn per day. This gives adequate chlorination for 10-15 days. Wells with higher draw-off rates need two chlorination pots.

The NEERI has also developed chlorine tablets and ampules or disinfecting water at the household level, particularly in rural areas. The health department of some state governments are stocking these tablets for free distribution in rural areas and those under epidemic. At present there are ten firms manufacturing these tablets and ampules.

For defluoridation of drinking water, NEERI, Nagpur, and Tamil Nadu Water Supply and Drainage Board, Madras, have developed two methods independently. Under the Nalgonda technique, an alkali (lime, soda ash or sodium bicarbonate), alum and bleaching powder is added to the water stored in a container. A tap is fixed at 3.5 cm from the bottom. Raw water to be used for drinking is put in the container and chemicals are added in adequate quantities. The water is surred for about ten minutes. The floc formed settles in an hour. The clear supernatant can then be drawn out by opening the tap. The sediment is discarded. This technique has been demonstrated in some villages of Andhra Pradesh, Tamil Nadu and Rajasthan.

For treating 50 litres of raw water the cost would be 2-10 paise depending upon the concentration of fluoride to be removed. The TWAD technique, developed by TWAD, Madras, can completely remove fluoride from the water. The plant is so designed that raw water pumped out of the well is conveyed through a pipe and doses of lime and alum are added to it in stages from containers kept above the water level. The water, mixed with these chemicals, then goes down to another pipe and enters from below into circular tank with a conical bottom filed with river pebbles. The treated water is tapped from top of the tank.

The important feature of the technology is that the flocs formed by the interaction of lime and alum with water settles down as a uniform layer above the pebbles instead of floating. This sludge blanket thus formed creates the proper conditions for the removal of fluoride. The TWAD technology is also successful in precipitating the aluminium derived from the alum. Thus, while improving the efficiency of fluoride removal, it helps in finally treating water that is also free of aluminium which otherwise would pose a health hazard if present in more than the permissible limit. The TWAD scientist have set up a pilot plant near Poonamalee, close to Madras, which has achieved 100 per cent success.

A number of techniques to remove salinity from brackish water making it potable has been developed by the Central Salt and Marine Chemical Research Institute (CSMCRI), Bhavnagar, and Bhabha Atomic Research Centre, Bombay, Desalination is now being accomplished in the country mainly by distillation processes, membrane processes, electrodialysis and solar distillation techniques. Work on solar distillation has been conducted at CSMCRI, National Physical Laboratory, New Delhi, and Defence Research Laboratory, Jodhpur. A 1,000 litres per day capacity was constructed by CSMCRI in Kutch.

NEERI, Nagpur, has also developed a test kit for measuring residual chlorine which is used either in the form of gas or as solution in water or as bleaching powder for disinfection of water supplies. Being very active it reacts with most of the impurities in water including disease causing micro-organisms. Thus residual chlorine in water indicates fulfilment of complete chlorine demand and the presence of excess for taking care of any later contamination. The kit helps in ensuring an optimum level of 0.2 ppm of chlorine in water to ensure complete disinfection.

During the sixth plan period over 1.92 lakh problem villages were given facilities for safe drinking water. In addition, 47,000 non-problem villages were also covered. However, figure of coverage of problem villages during the sixth plan period appears prima facie impressive, but it may be mentioned that substantial portion of the coverage was only partial in future.

To eliminate the drudgery involved in making 'donas', a simple machine to make cups or plates from various kinds of leaves has been developed by CFTRI, Mysore. Various models of improved chullhas (stoves) have been developed. They could be made with locally available mud mortar, cattle dung, sand straw, red clay, A.C.

pipes, metallic damper and metallic hool. The new chulhas are 30 per cent more efficient than the traditional ones.

The RRL, Jorhat, has developed a paper slate which proved a good substitute for the conventional costly stone slates. The paper slates are light, unbreakable, cheap and do not lose abrasiveness quickly. The raw materials required for making the slates are paper board and some chemicals.

Several indoor and outdoor storage bins have been developed and tested by many institutions in India including Central Building Research Institute (CBRI), Roorkee, Structural Engineering Research Centre (SERC), Ghaziabad, Indian Plywood Research Institute, Bangalore, and Indian Grain Storage Institute, Hapur (U.P.). These bins could be made of mud, bamboo, earthenware, masonry metal and wood individually or in combination.

The CFTRI, Mysore, has also developed following processes to control pest: (1) durofume process for protecting commercial stock of food grains; (2) pest proofing of jute gunny bags used for storage of food grains by means of a portable machine (400 bags/hr; (3) low-cost fumigant tablets called 'Minifume' (one tablet costing about Re. 0.60 is enough to treat 100 kg of foodgrains); (4) an integrated technique of eradication of domestic rats involving fumigation of rat burrows and combined use of repellants, attractants and poisoned baits.

The RRL, Bhubaneshwar, has developed a 'food preservation flask', in which cooked food can be preserved for a period of three days without refrigeration.

The efficiency of the traditional bullock cart is low and despite this it is being used as a means of rural transport in the country. Many new designs have been developed to increase its efficiency and reduce the strain on the animals. Major innovations in the new bullock carts are the hand operated shoe-type brakes with spokes for increased load carrying capacity and bearing, for smooth running. Two such stub-shafts eliminate the axle shaft to facilitate easy turning.

About 26 research institutions in India have evolved more than 52 new construction techniques and materials which can reduce the cost of construction as well as conserve scarce and costly inputs like cement and steel.

A number of new construction techniques have already been incorporated by the National Building Organisation (NBO) in its experimental housing project. They are also being adopted by various construction agencies in the country. These include single brick thick load bearing walls for up to five storeyed residential buildings, under-reamed pile foundation, precast RCC roofing system and precast lintels.

The Central Building Research Institute (CBRI), Roorkee, has developed a technique known as under-reamed piles for laying foundations in black cotton soil areas whose soil bearing capacity is low and subject to alternate cycles of swelling and shrinkage. It has developed a technique for making improved roofs from thatch. The thatch could be made of paddy, straw, coconut leaves of palmyra leaves of reeds. The life is four to five years, as against about a year of the traditional made thatched roof. The institute has also developed a non-erodable plaster for protecting mud walls against rains. It would greatly help villagers in reducing their house maintenance costs. Recently RRL, Trivandrum, has developed a mixture of urea and diammonium hydrogen phosphate which proved to be a good fire retardant for coconut leaf thatch used as a roofing material in the coastal areas. It has also been found that a combination of copper sulphate and phosphorylated cashewnut shell liquid not only enhance life of coconut leaf thatch but also provides fire retardant property. Aluminium phosphate has also been found to impart fire retardancy to thatch.

The CBRI and RRL, Jorhat, have developed a process for manufacture of cheap roofing made of paper sheets which are light, strong, fire resistant, non-absorbant and unbreakable. It is easy to transport and install.

The construction of under-reamed piles is very simple and it requires equipment like auger boring guide, earth auger, under-reams, wheel barrow and concrete funnel. These are board concrete piles having one or more enlarged base towards the bottom. The principle involved is to another the building by means of these piles at a depth where the soil has a nearly stable moisture content during the various seasons of the year.

These piles can be successfully used for machine and factory foundations and foundations of transmission towers. The piles have been standardised to avoid inconvenience in the construction. The main diameter of the piles ranges from 20 cm to 45 cm and the length from 3 m to 8 m. The diameter of the under-ream bulb is generally two and a half times that of the piles. The minimum spacing of the under-ream is 1.5 times, the diameter of the under-ream itself. A reinforcement case is also provided through the length of the pile.

Under the single-brick thick load bearing walls, techniques which use better quality of bricks, and improved methods of construction, the thickness of walls of four storeyed buildings could be significantly reduced resulting in economy in cost and consumption of materials.

Appropriate materials for roofs which contribute 30 to 40 per cent of the total cost of construction have been developed by various institutes in the country. The Regional Research Laboratory, Trivandrum, has developed a process which extend the life of coconut leaf thatch from one to four years. A separate process has also been developed at the laboratory which makes coconut leaf thatch fire-retardant for short term uses. It has been found that a combination of copper sulphate and phosphorylated cashewnut shell liquid (CNSL) not only enhances life of coconut leaf thatch but also provides a fire-retardant property. Aluminium phosphate has also been found to impart fire retardancy to thatch.

The CBRI has developed good quality clay tiles which increase the life of roof. The use of cement and steel can be avoided by making the roof with the clay tiles. Methods to manufacture asbestos or G.O. or aluminium corrugated sheet which are light in weight and economical in cost, have been developed. But the disadvantage of such roofs are that these get heated up quickly. Asbestos sheets are brittle and these get rusted in humid areas. The CBRI has also developed cement-coir-corrugated sheets for roofing which do not suffer from these defects.

The NBO has developed a method for manufacturing of corrugated sheets from waste materials like scrap paper, bagasses, jute wastes and coconut fibres. These sheets have good thermal insulation properties and are water proof, flexible and light in weight. They are not susceptible to corrosion and leakage and are relatively safe against fire and are not brittle like asbestos sheets. In this type of roofing use of cement and steel is practically nill. It has been recommended to be used in the areas where the temperature does not exceed 40°C.

For flat roofs, several systems like RC channel units, RC plank and joist system, doubly curved tiled roofing, precast circular units, L panel roofing, cord unit system, precast wall, shell unit and battens and hollow block system have been developed by various institutes which cut down consumption of steel and cement ranging from 5 to 20 and 15 to 30 per cent respectively.

The CBRI and RRL, Jorhat, have developed a process for the manufacture of cheap paper roofing sheets which are light, fire resistant, low waste absorption, high strength, no breakage in transport and easy to install. These sheets are comparatively much better as against the commercially available one.

In search for substitute of cement, the Karnataka State Council for Science and Technology (KSCST), Bangalore, has found that starch solution and jaggery syrup could be substituted for the costly cement and lime for construction purposes. The jaggery syrup, prepared by boiling waste jaggery with water till it acquires stickly consistency, can be mixed with red soil and then pressed into solid blocks. Similarly starch solution prepared by baking maida flour in water can be pressed into blocks. These blocks were found to be erosion resistant after thorough drying. It has also developed prefabricated light weight roof panels using bamboo, paddy husk, soil and cement.

A process of hydration of lime has been developed which makes available hydrated lime in dry form, standard quality and ready to use condition as in the case of cement. It can be stored in bags for 6 months without any deterioration. It can be used either alone or in combination with cement or clay pozzolana.

A new process of production of clay pozzolana produced by grinding brickbats has been developed in which clays of suitable specifications are calcinated to the optimum temperature and then grind to fine powder as required by the Indian Standard Specifications. The processes have been developed for production of clay pozzolana by using down draft kiln by the Central Road Research Institute.

In order to improve quality of timber, the Forest Research Institute, Dehradun, has developed techniques for seasoning of timber which is essentially a drying process. It decreases weight and generally increases strength and helps in retension of shape.

Investigations carried out in Research Laboratories have shown that flyash, a fine residue obtained from thermal power stations, can be utilized in various form as building material. It can replace cement 20 per cent by weight without adversely affecting the strength of concrete.

The College of Military Engineering, Pune, has fabricated houses which essentially consists of prefabricated RCC frame work, precast ferrocement, roof panels, half-brick thick walls and brick paved floor. Cost of construction comes to about Rs.300 per sq metre.

The CBRI has also conducted intensive studies on planning, design and construction aspects of the houses during the last four decades with a view to obtaining optimum utilization of space, increase in functional efficiency and reduction in cost of construction. Based on these studies a single storey dwelling unit for a family of our members has been evolved.

The design is said to be structurally sound, functionally efficient, aesthetically appealing and offer an economy of about 10 per cent in the overall cost. The plinth area of the house is 48.6 sq. m. The estimated cost of construction of the house was around Rs. 45,000.

Under the experimental housing scheme, NBO has tried out new buildings material and techniques in 42 projects costing Rs. 30 million in different parts of the country. The techniques and materials tried out were under reamed pile foundation, precast RCC roofing system, thinner load bearing brick walls and use of high strength deformed bars.

Mushroom Culture

The ICAR and RRL, Jorhat, have improved a method for cultivation of edible mushroom. Under the method paddy straw is cut into pieces and soaked in water for about two days. The straw is drained and filled in polythene bags. After incubating in bags in a room, mycelium is innoculated and the bags are kept in a properly aerated room. The bags are regularly irrigated and the mushrooms start appearing after about five days.

A simple high-efficiency wood burning stove has been developed by the Central Power Research Institute, Bangalore. The stove consists of a cylindrical chamber closed at top and the bottom with slotted plates of specially evolved pattern a grate is fitted inside the chamber slightly above the bottom slotted plate.

The dimensions of the plate and the chamber have been carefully worked out to achieve maximum combustion. An aluminium sheet linings provided in the chamber to reflect heat towards the centre of the combustion chamber and to minimise heat loss through radiation from the walls. These design feataures have boosted its thermal efficiency to over 30 per cent.

Farm Machinery

In the area of farm machinery, several useful machines and implements have been produced including bullock drawn sweeps, chisel plough puddler, manual and animal drawn seed-cum-fertilizer drill, power tiller operated paddy harvestor, tractor oriented reaper binder and multicrop thrasher. Besides, several tools have been further improved namely hand holes, ploughs horrows and cultivators. In the case of processing equipment several innovations have taken place. Mention can be made of equipment like grain dryers for cereal crops, groundnut decorticator, sun-flower and groundnut cracker. Several designs of cowdung and biogas plants have been developed and single technologies have worked out to recycle agricultural wastes.

Several models of solar cookers including simple hot box type solar cookers with the reflector, solar oven with multiple reflectors and solar concentrators have been developed under a scheme, sponsored by Commission for Additional Sources of Energy. Under the demonstration and field trials programme in this area, systems like water heater and dryers, wood seasoning systems and pumps are being installed.

Besides development of appropriate technologies to rural masses, a number of schemes relating to the transfer of relevant technologies for landless labourers, artisans, small and marginal farmers and tribal communities like setting up of krishi vigyan kendras, the lab to land agricultural programmes of the ICAR, polytechnology transfer centres of the CSIR, training of rural youth for self employment (TRYSEM) of the Department of Rural Reconstruction and Health for all programmes of the ICMR are being implemented. The NRDC and KVIC have been able to take the relevant technologies to rural areas with the help of its regional offices or centres located all over the country.

The Indian Jute Industries Research Association, Calcutta, has developed a technology for suitable binding jute / wool and jute / viscose to make woollens.

The Central Electro-Chemical Research Institute (CECRI), Karaikudi, has made a breakthrough in dry cell technology using indigenous raw materials and machinery. The cell produces are comparable to other cells marketed in the country.

The RRL, Jammu, has developed a process to produce fibre boards from pine needles. Packing cases made of these boards have been found useful for packing and transporting fruits over long distances. These packings are cheap and durable and easy to assemble or dismantle.

Although over 4450 rural technologies developed in the country have been adopted by villagers, there is more to be done to make innovations and inventions a way of life in rural areas.

Rural Roads

The rural road building engineers have progressively improved and standardised various techniques of building roads in rural areas and are applying them to field conditions. At least six economical and efficient techniques of soil stabilisation have been standardised by engineers of the Central Road Research Institute (CRRI), New Delhi. Many handy and energy-saving tools have also been designed and fabricated by them.

Besides there are more than 30 institutions, big and small, working in this direction and have achieved varying degrees of success and sophistication of approach and technologies. The Central Road Research Institute is the largest and the best equipped organisation with rich experience and achievements behind it.

Before constructing a rural road, the soil on which it rests has to be treated and stabilised so that potholes do not appear and spoil the surface. The surface may also crack on unstabilised soil, making movement of vehicles virtually impossible. CRRI engineers have standardised six techniques of soil stabilisation, namely, mechanical, lime, cement, lime-flash, bitumen and two stage stabilisation.

These techniques vary according to soil conditions and materials available locally because cost is the main constraint in rural roads construction. Mechanical stabilisation is applied to soils containing sand, moorum / gravel with missing fractions and clayey soils. Missing fractions are blended (e.g. clay with sand) so as to produce a mass of maximum possible density end with plasticity within limits. Too much plasticity is not desirable— it makes the subgrade weak.

Lime stabilisation is the most preferred treatment for its free availability and cheap price. Black cotton soil in Maharashtra, Andhra Pradesh, Madhya Pradesh and Uttar Pradesh poses a problem for road construction as the soil shrinks and cracks during dry shells. This would crack the road surface as well. But if only five per cent of lime is added to the soil, a chemical reaction takes place, and the soil becomes ready for laying roads, CRRI engineers are using this technique for stabilising soils in these states. This works out to be very cheap and lasting. For this purpose, lime is dehydrated from reacts with clay minerals in the soil to cause immediate reduction in plasticity because of cationic exchange, flocculation and agglomeration. Lime is also used for long term chemical reaction with clay minerals to produce cementitious products which bind the soil for increased strength and stability.

In the cement stabilisation techniques hydrated cement products bind soil particles, the strength developing depending on the concentration of cement and the particles. But, cement treatment of soil is everywhere, until it becomes unavoidable, such as in the case of granular soil without high concentrations of organic matter.

Clayey soils and soils with medium plasticity are stabilized by the lime-flyash stabilisation techniques. Lime reacts chemically with silica and alumina in flyash to form cementitious compounds which bind the soil. Bitumen stabilisation is not the preferred mode of stabilisation for its excessively prohibitive cost factor although it is settled beyond doubt that it is a superior material making the road impervious to water and many environmental factors.

Soil stabilisation forms the foundation of road building and thus has a special place in the technology. Road engineers have very skillfully and economically used locally available materials to stabilise soil for Indian village roads, Kankar (small pebbles), moorum, laterite, dhandhala, gravel, lime and other local materials mixed in right proportions with soil have been found to stabilise soil very efficiently, especially since the roads are meant for bullock carts with damaging solid wheels.

While building roads for solid-wheeled carts, the fact that has to be kept in mind is the destructive nature of cart's wheels. Because of very localised stress and pressure on the road and sharp-edgeness of the steel lined wheel, deep tracks develop on the surface and finally the road given way. For this purpose, the thickness of the material applied has to be appropriate to withstand the onslaught. CRRI engineers have worked out a formula which will tell exactly what should be the thickness of the material used.

The contact area of the steel-rimmed tyre on the road depends on the extent of wheel load, width of tyre, road camber, degree of self alignment of wheel, axle system, and the flexibility of degree of yielding of the surface. Based on these parameters CRRI engineers have plotted a design curve which says that for most of the soil, the required pavement thickness should be around 9 inches. This is being applied throughout the country with successful results.

Pavement thickness being a very important consideration in road building, CRRI technololgists have worked out an index which indicates the exact thickness of the pavement required depending on the stress and load bearing capacity of the soil and materials used. It is a very straightforward curve which would immediately tell the engineers what should be the thickness of the pavement to carry a given set of vehicles on a particular soil type. According to the stress, bearing capacity of the soil, the additives used to stabilise it, the upper layers of the roads and the types and average number of vehicles plying the route it is the decided what the thickness should be. Soil and subgrade compaction are also important factors in rural road construction. The more compacted the material the better its ability to withstand stress and load. CRRI scientists have standardised methods of soil compaction on the basis of materials, soil, equipment used and the pavement thickness required.

Building rural roads is a necessity but it should be with an eye on economy and usefulness. CRRI scientists, after studying dozens of villages and roads connecting them, their commercial status, agricultural development, proximity to market, highways and main roads and towns, as well as their topography have designed on overall approach to building roads for their use which would best serve their interest and be economical at the same time. CRRI supplies precise information on how to connect a village with a township or a market centre and how to route it.

Such models are of great use in preparing master plans for rural development at the national level, taking the district as the basic unit.

Quality control being essential for ensuring proper construction and serviceability of roads, CRRI has developed a wide range of easy, quick and efficient tests that can be conducted on site. The tests, like evaluation of soil type and mixture content, are simple enough to be handled by an undergraduate with preliminary training in handling the equipment.

CRRI engineers have also designed precast concrete paths for rural roads in which two parallel running concrete pavements are laid to provide a hardy surface for carts to ply.

The institute is conducting tests on the quantitative effects of various compositional factors on the strength and deformation characteristics of granular mixes. This study is important as the requirements of bitumen, a petroleum based material, have to be cut down for economy in road building. Equipment and test procedures have been developed and devised making possible the comparative evaluations of granular mixes under controlled conditions.

In Western Rajasthan besides the fine dune, sand and the kankar / dhandhala here and there conventional construction material as well as water are not generally available locally. Though initial design and successive trials under static and repetitive loading, design has been optimised for medium intensity of traffic. The design consists of ribbed hexagonal precast concrete blocks with convenient arrangements for interconnections and load transfer.

Improvements in bullock carts have also been effected at the CRRI. The institute developed a wheel tester for testing wheel designs, wheel axle system and bearings. A procedure was developed for the assessment of contact stresses on the pavement due to solid wheels and concepts were developed for pavement designs based on solid wheel traffic in rural areas.

In more recent years attention has been given to the development of low cost bearings as well as flexible element which improve the contact between the wheel and the travelled surface of changing cross profile and this would reduce the tracking effort.

Under a sponsored scheme of the Ministry of Transport, bullock cart prototypes with innovative energy-saving designs were fabricated and distributed to farmers in and around Delhi and in the northern states of the country for actual field trials, and as a result some improvements have been made in cart design.

Landslides pose a grave danger for road builders as well as for road users. The evaluation and testing of landslide prone areas and soils is regularly done and many instruments have been indigenously fabricated for such tests. Some of the instruments so developed are the composite differential settlement-cum-tilt meter and the remote control hydraulic settlement gauge. Other instruments which have been developed for this purpose relate to the fabrication of a slope indicator for monitoring small ground motions in earthquake projects and hillside rocks.

In the process of evolving suitable and economical specifications for rural roads CRRI scientists found through laboratory research and field experimentation that soft aggregates like brick ballast, kankar, dhandhala coral stone could be used instead of costlier stone aggregates for rural roads.

Keeping in view the non-availability of power driven road rollers in villages, CRRI engineers developed easily fabricated and ferried road building devices. A collapsible animal drawn road-roller so developed comes in very handy. It consists of a wooden box mounted on the metallic roller. The box is empty but when filled with rocks or any other heavy material works as a light roller and is drawn by an animal. Another roller is even lighter. It is made of a hollow but thick-bodied drum. When filled with rock it works as a roller. It certainly takes more rounds than a power roller would, but there are thousands of villages especially in hilly and marshy areas where power driven rollers just cannot reach. Here such rollers can do the job just as efficiently, and are far cheaper.

A water bowser designed and fabricated by CRRI engineers is actually an improvised device. Five to six big metal drums are mounted atop a cart. They are connected together by means of pipes. A perforated pipe at the rear end of the cart sprinkles water on the road while it is under treatment and rolling. It is very cheap and is used where mechanical or power driven pumps are not available.

CRRI scientists have also developed and fabricated an animal drawn aggregate spreader for roads. It is part manual. It spreads the aggregate mix uniformly and works much faster than man but not the mechanical spreader.

Road building technology for rural areas is fairly improved in India and future developments will strengthen it further. But a nagging issue is the limited number of villages in the country which have serviceable all wheather roads. More than seventy per cent of the over 780 million Indians live in 600,000 villages dotting the entire Indian landmass, but only one-third of the villages have been provided with all weather roads. Estimates are that Rs. 25,000 million will have to be spent to give every village an all weather road by the turn of the century.

The avowed aim of the Government to construct 27,00,000 km of roads by 2000 AD calls for a massive organisational structure and the easy availability of funds, materials and expertise. About 10,00,000 km of rural roads need to be built to let nearly 15 million animal drawn carts in the country ply and carry most of the grain and material and ferry cattle and people from one place to another.

Environment Management

In the field of environment management, Indian scientists have developed facilities for management of radio-atomic wastes and chemical effluents of atomic power plants, reclaimed waste water, and supplied cheaper piped gas by recycling urban wastes. They have also generated electricity from municipal wastes and undertaken the gigantic task to clean the Ganga, the symbol of India's age-long culture and civilisation. The main thrust of the Ganga action plan is to purify the river to prevent the discharge of effluents with a high pollution load by renovation of all existing sewage pumping and treatment plants and installation of new plants.

Works in this area has been on air quality monitoring and pollution control, water chemistry and treatment, solid waste management, rural sanitation and epidemiology. The Department of Environment (DOE) lays major emphasis on environmental and ecological research, environmental impact assessment, monitoring of environmental quality, setting up of environmental information system, restoration of fragile ecosystems, preparation of an inventory of ecological resources, soil and water conservation, forest conservation and aforestation, social forestry, economic and industrial plantations, energy plantations, drinking water supply and sanitation.

Waste water reclamation through aqua-culture, optimum utilisation of domestic and industrial waste water in agriculture, a detoxification of phenol and cyanide bearing waste by biological methods have been the major areas of emphasis at the National Environmental Engineering Research Institute (NEERI), Nagpur. Studies on seweage and industrial waste treatment and its utilisation have been taken up by the NEERI for over 250 industries manufacturing synthetic drugs, pharmaceuticals, pulp and paper, straw board, rayon, chemicals, high explosive and fertilisers. Designs for appropriate mechanical compositing plants have been developed by NEERI after careful characterisation of solid wastes of over fifty major cities.

One such large plant for urban waste recycling is already existing in the country and is being operated successfully at Okhla near Delhi. The plant comprises 15 digestors connected to gas collectors. The total gas generation from the plant is about 0.6 million cubic feet per day, having a heat value of 700-800 BTU per cubic feet. The gas is being successfully piped to about 800 households spread over an area of four kilometers. The cost of investment in sludge gas is about 50 per cent cheaper than the ordinary LPG cylinder gas.

Based on the successful experience and considering the potential of sewage sludge treatment systems as a renewable energy source the government has assigned due importance to this area. One such project has been commissioned at Padrauna in Uttar Pradesh, and plants are under construction at Ayodhya in Uttar Pradesh, Keshaopur (Delhi) and Bhopal (Madhya Pradesh). Experience from these pilot projects is expected to provide a basis for extending the activities to other potential cities in the country.

The combustion of refuse solids converts the combustible components to inert substance generating heat and resulting in 75 per cent reduction in volume. The process is known as incineration. About 25-45 per cent of heat becomes surplus after supporting combustion and could be utilised as a source of energy. Though conventional incineration is the usual method adopted for disposal of urban wastes elsewhere, only about 13 per cent of the total urban waste is processed using this technique. The efficiency and output of the incineration is, however, low and presents problems like requirement of large incinerations, firing devices (refuse being voluminious) requirement of storage facilities which may lead to purification of organic matter and air pollution in the absence of proper control devices. Viable technologies have merged in the country for successful incinerators. In Delhi, the first pilot incineration plant which will process 300 metric tonnes of municipal solid waste per day of power generation has already started working. Such plants are likely to be set up in other cities like Calcutta and Madras.

Pyrolysis—chemical decomposition of carbonaceous material by that in the absence or in the controlled oxygen atmosphere—is another promising indigenous technology for resources recovery from solid waste. The process has the advantage of reducing the volume of waste to 80-90 per cent with minimum chances for air pollution and requires less land area. The product of pyrolysis and indirect gasification are fuel gas, mainly carbon monoxide, hydrogen, methane, carbon dioxide and charcoal.

Delhi and Maharashtra have taken up setting up of pilot plants based on the pyrolysis gasification technology for treating solid wastes. Perfection of this technology in Indian conditions and success of these pilot plants may open promising avenues for resource recovery from urban waste in the country.

A plant for detoxification of phenols through soil microbial cultures of an LTC plant in Andhra Pradesh has been set up by NEERI. Four slow sand filtration plants have already been set up based on know-how developed by NEERI.

While an anaerobic digestion technique using a special filter has been developed at the Central Leather Research Institute, Madras, to reduce the BOD content of tannery effluent, the National Institute of Oceanography, Goa, has found out suitable discharge point for treated effluents from portbased steel plants and fertiliser plant.

Epidemiological surveys have been conducted by the Industrial Toxicological Research Centre, Lucknow, to determine the health status of the working population to unravel existing as well as hitherto undetected health hazards to workers in specific industries exposed to cyanide fumes, lead poisioning in jewellery manufacture. Two bacterial strains for microbial degradation of plastics and other material used in the polymer industry have also been isolated by the ITRC.

The NEERI has taken up major research projects on analysis, assessment and chemical treatment of toxic discharges from phosphatic fertiliser and basic organic chemical industries, assessment of utilising effluents from copper, zinc and fertiliser plants and refineries and agricultural usage and impact of fugitive and stack emissions from selected industries such as fertilisers and cement plants on neighbourhood air quality.

Programmes on health problems arising out of exposures to environmental pollutants in industries especially synthetic dyes, pesticides, plastics, petrochemicals, metals and minerals are being undertaken by ITRC. Standards for thermal stress on industrial workers in an industrial environment are being evaluated at the Central Building Research Institute, Roorkee.

The Bhabha Atomic Research Centre, Trombay, has completed construction of solid state surveillance facility at Tarapur for the interim storage of solidified high level radioactive waste packages produced in the Waste Immobilisation Plant (WIP) at Tarapur. Facilities for management of radioactive wastes and chemical effluents at Tarapur, Kota, Trombay, Hyderabad, Alwaye and Kalpakkam are functioning satisfactorily. Experimental work is under way to estimate feasibility of deep geological formations for radioactive waste disposal.

While the Cotton Technological Research Laboratory (CTRL) in Bombay has designed and fabricated a biogas plant in which willow dust, waste material generated in textile units, is used as raw materials, the Central Building Research Institute, Roorkee, has developed a system to dispose waste water near the source point. The system consists of two chambers — an ash / silt tap chamber and filtration chamber. The first one is meant for water in which silt or ash will settle down and the other filled with brick aggregate for removal of the suspended or colloidal matter in the water. The drain waste so filtered flows into a bore hole for ultimate disposal.

A study conducted by the Central Soil Salinity Research Institute, Karnal, has revealed that use of untreated sewage water for irrigating vegetables is not good for health as vegetables may contain pathogens and toxic elements like lead, nickle, mercury and copper.

The potential of flora and fauna as indicators and containers of pollution, jointly taken up by ITRC and the National Botanical Research Institute, Lucknow, is being evaluated. Alcohol was extracted by the scientists of Indian Institute of Technology, Delhi, from distillary wastes.

Use of tannery effluents to grow water hyacinth for investigating the possibility of biodegrading toxic elements is an important on-going project at the Central Leather Research Institute, Madras.

With the creation of the Department of Environment in 1980, the Government of India has pursued nature conservation and management on a broader scale, than the time when the National Committee on Environmental Planning and Coordination was serviced by a small division in the Deparatment of Science & Technology. This national level committee, created in 1972, was preceded by a committee on human environment set up in 1970.

The Department of Environment's functions cover wide ground, including coordination, policy planning, regulatory functions, conservation and promotion. To ensure effective performance by the department complementary work in several areas is being pursued. These include pollution control, environmental assessment, living natural resources, conservation, eco-development, environmental research promotion, environmental training, education and awareness of environmental law, society, environment interactions, coordination/liaison with state governments, environmental information, international cooperation, administration and finance.

Some legislations that are being enforced in the country are as follows : The Water Act (1974), the Water Cess Act (1978), the Air Act (1980) and the Wildlife (Protection) Act 1972. Recently the government has enacted the Environment (Protection) Act 1986 to check pollution and conserve flora and fauna. It empowers any citizen to seek legal recourse against parties harming the environment.

The department has been working closely with some organisations associated with wildlife conservation. These include the Indian Board of Wildlife, the Zoological Survey of India, the Botancial Survey of India, the Indian Institute of Wildlife Management and the Delhi Zoological Park.

NEW CONSTRUCTION TECHNIQUES AND DESIGN CONCEPTS EVOLVED BY VARIOUS RESEARCH INSTITUTIONS

1. Central Building Research Institute (CBRI), Roorkee:
   - (i) Under-reamed pile foundation
   - (ii) Brick-on-edge cavity walls
   - (iii) Single thickness load bearing brick walls for 4 and 5 storeyed buildings.
   - (iv) Improved method of brick laying and plastering
   - (v) Polythylene damp proof coarse
   - (vi) Stone block masonry
   - (vii) Light weight aggregate concrete blocks for partitions
   - (viii) Precast RC channel unit for roofing/flooring
   - (ix) Precast RC cored units for roofing/flooring
   - (x) Precast RC cellular units for roofing/flooring
   - (xi) Precast RC waffle units for roofing/flooring
   - (xii) Doubly curved tiles roofing system
   - (xiii) Precast RC solid planks for roofing/flooring
   - (xiv) Precast RCL panel roofing system
   - (xv) Precast reinforced brick panels for walls and roofing/flooring
   - (xvi) Particle board for door shutters
   - (xvii) Precast RCC frames for doors and windows

(xviii) Magnesium oxychloride door frames

(xix) Magnesium oxychloride tiles for flooring

(xx) Stabilised soil cement base for cement concrete floor

(xxi) Partial replacement of cement by flash in mortar and concrete

(xxii) Single stack system of plumbing

(xxiii) Precast thin lintels

(xxiv) Water proofing in roof with lime concrete terracing

(xxv) Tamping machine for tamping lime concrete on roof

(xxvi) Surface water proofing of mud walls

(xxvii) Corrugated roofing sheets from coir waste

(xxviii) Fire retardant treatment of thatch

(xxix) Improved clay roofing tiles

(xxx) Improved design of brick kiln and lime kiln

2. Structural Engineering Research Centre (SERC), Madras:

(i) Hyperbolic paraboloid footings

(ii) Precast RC waffle unit for roofing/flooring

(iii) Prestressed concrete hyperboloid shell

(iv) Precast doubly curved shells for roofing/flooring

(v) Hyperbolic paraboloid shell roof

(vi) Brick shell roof with flat bricks

(vii) Lime-Flyash cellular slab for roofing

(viii) Use of high strength deformed bars and ultimate design procedures

(ix) Use of flyash in mortars and concrete

(x) Use of flyash in precast prestressed concrete products

(xi) Ferro-cement concrete water tank

(xii) Ferro-cement concrete silos, bins, etc., for grain storage

(xiii) Precast slab, using hollow clay blocks for floor roofs system

(xiv) Precast RC joists and hollow clay blocks roofs

(xv) Precast channel units

(xvi) On site large panel prefabrication

(xvii) Precast two-way spanning flooring / roofing scheme

(xviii) Electrothermal prestressing

(xix) Prestressed concrete railway sleepers

(xx) Fibre reinforced concrete manhole covers

(xxi) Prestressed concrete poles

(xxii) Latro blocks (building blocks from latrite soil)

3. National Council for Cement and Building Materials, Delhi

(i) Use of flyash in mortars and plasters

(ii) Use of flyash in concrete

(iii) Production of ready-mixed concrete

(iv) Production of strength mix design of concrete

(v) Cement soil stabilised blocks

(vi) Precast roofing tiles for low cost housing

(vii) Portland pozzolana cement

(viii) Prevention of chemical attack on reinforced concrete structures

4. Central Road Research, Institute, Delhi:

(i) Reactive Surkhi

(ii) Waterproof mud plaster on mud walls

(iii) Flyash as partial replacement of fine aggregate in cement concrete

(iv) Lime / burnt clay pozzolana

(v) Direct field adjustment chart for control of concrete quality

(vi) Bituminous materials in the water proofing of mud walls

5. Central Electro-Chemical Research Institute, Karaikudi, Tamil Nadu:

(i) Prevention of corrosion of reinforcement in brick work

(ii) Prevention of corrosion of steel reinforcement cement concrete

(iii) Prevention of corrosion of steel

(iv) Corrosion in buildings—level of corrosibility in different regions of India

6. Forest Research Institute, Dehradun:

(i) Secondary species of timber

(ii) Small and medium span nail jointed timber trusses

(iii) Utilization of short length small dimension timbers for construction purposes

(iv) Termite proofing of building and houses

(v) Fire retarding and preservative treatment of thatch roof

(vi) Preservative treatment of bamboos

7. Indian Plywood Industries Research Institute, Bangalore:

(i) Veneers plywood shingles for roofing

(ii) Exterior grade plywood for roofing and walling

(iii) Hard-board for exterior applications

(iv) Glued wood - plywood structural components like beams and structural components

(v) Glued laminated doors and window frames and other glued laminated products using structural synthetic resin adhesive

8. Regional Research Laboratory, Jorhat:

(i) Cement bounded fibre roofing sheets

(ii) Rice-husk flyash pozzolana

(iii) Paddy husk bricks

(iv) Paper corrugated roofing sheets

(v) Flooring tiles for A.C. waste

(vi) Bamboocrete for wall panel and roof

(vii) Water filter candle for obtaining drinking water

(viii) Cheap sanitary septic tanks from waste bitumen drums

9. School of Research & Training in Earthquake Engineering, University of Roorkee:

(i) Single brick thick load bearing walls for 4 storeyed residential buildings

(ii) Single brick thick load bearing walls for 5 storeyed residential buildings

(iii) Half brick thick 'Z' shaped load bearing walls

(iv) Seismic design of present RC systems for roofing / flooring

10. Hindustan Prefab Ltd., Delhi:

(i) Partial prefabricated framed structure with RC columns and beams

(ii) Pocket connections for precast columns with foundation

(iii) Load bearing concrete panels for walls

(iv) Hollow cinder block masonry

(v) Lime and flyash hollow block masonry

(vi) Precast and battens and hollow blocks for roofs and floors

(vii) Precast RC cored roofing / flooring system

(viii) Prestressed cement concrete beams

(ix) Prestressed concrete hollow covered slab resting in prestressed concrete beams for roofing and flooring

(x) Precast RC single flight staircase monolithically cast with landing

(xi) Prestressed concrete frames for doors and windows

(xii) Modified horizontal and vertical joints between concrete wall panels

(xiii) Full panel prefab method of construction

(xiv) Channel slab for roofing in residential and industrial buildings

11. Bureau of Indian Standards, Delhi:
   (i) National Building Code
   (ii) Standards for low-income group housing
   (iii) Modular coordination in buildings
   (iv) Standard building specifications and codes
   (v) Standard method of measurements
   (vi) Standard output norms for materials

12. U.P. PWD Research Institute, Lucknow:
   (i) Strength of brick masonry and mortars
   (ii) Waterproofing of flat roofs
   (iii) Replacement of cement by surkhi
   (iv) Damp proof construction of walls

13. Planning, Research and Action Institute, Lucknow:
   (i) Investigations of comparative efficiency of different types of chullhas and evolving design of smokeless chullhas

14. Sri Ram Institute of Industrial Research, Delhi :
   (i) Production of clay pozzolana by fluidized method
   (ii) Production of gypsum plaster by fluidized method

15. Building & Road Research Laboratory, Chandigarh:
   (i) Bitumen stabilised soil bricks

16. Central Fuel Research Institute, Dhanbad
   (i) Flyash bricks

17. Central Mechanical Engineering Research Institute, Durgapur:
   (i) Development of simple brick making machine
   (ii) Economized steel structure of open web sections
   (iii) Design development for castellated beams

18. National Environment Engg. Research Institute, Nagpur:
    - (i) Sanitary rural latrines
    - (ii) Garbage chutes in multi-storeyed buildings
19. College of Military Engineering (CME), Pune:
    - (i) Precast roofing/flooring units such as cored units, channel units, waffles shells and hyperboloid shells
    - (ii) Damp proof course in cement mortar (1:4) with air entraining agent
20. Regional Engineering College, Warrangal:
    - (i) Composite joists filler blocks for roof floors
    - (ii) Ribbed slab floor roof (large units)
    - (iii) Ribbed slab floor roof (small units)
21. Bengal Engineering College, Howrah (West Bengal):
    - (i) Precast RC columns with bracket erected and fixed in recesses of RCC footings laid in situ
    - (ii) Precast and prestressed beams resting on brackets of the columns
    - (iii) Composite precast RC battens and hollow cinder blocks for flooring
    - (iv) Precast and prestressed folded plate roof
    - (v) Hollow cinder blocks masonry
22. Maharashtra Engineering Research Institute, Nasik:
    - (i) Use of surkhi in cement mortar as an inert
    - (ii) Durability of pozzolana mortars and concrete
    - (iii) Unburnt building bricks
23. I.I.T., Madras:
    - (i) RC portal frames construction
    - (ii) Clay products for roofing

(iii) Composite slab with RCC joists and brick in fill for floors / roofs

(iv) Hollow grid slabs for roof / floors

(vi) Simple hand operated hollow block making machine

(vi) Shell foundation

24. I.I.T., Kanpur:

(i) Cement from rice-husk

25. Neyveli Lignite Corporation, New Delhi:

(i) Flyash lime gypsum brick for masonry

26. National Building Organisation, Delhi:

(i) Modudlar concept in planning and design of building

(ii) Clay pozzolana for mortars

(iii) Dry-hydrated lime for mortars and concrete

(iv) Cellular concrete building products

(v) High strength bricks and structural clay products

(vi) Asphaltic roofing, sheets

(vii) Ribbed floors / roofs of concrete

(viii) Load bearing 19 cm thick brick walls

(ix) PVC pipes

(x) Plastic stays for steel windows

(xi) Hundson brick making machine

(xii) Sundried bricks and non-erodable water proof mud plaster

(xiii) Fire retardant treatment of thatch

(xiv) Preservative treatment of bamboos

(xv) Reed boards for roof and walling

(xvi) Bamboo reinforced concrete for roofing

(xvii) Lower ceiling height

14

MACHINE TOOLS AND ROBOTICS

The technological revolution taking place in the area of machine tools, inspection devices and handling equipment is now poised to bring rich dividends both in productivity and quality, the two factors required for success in an increasingly competitive market.

The new revolution has been triggered off by electronics and sustained by ever increasing capabilities of computers. This has led to emergence a new technology called 'Mechtronics' symbolising the synthesis of mechanical aspects. The heart of mechtronics revolution, Computer Numerical Control, more popularly known as CNC technology, made a slow start in the 1960s in the world, its phenomenal growth was witnessed only in 1970s and 1980s by virtue of its contribution towards higher productivity and substantial cost reduction.

In India, the first step before initiating in-house R&D was to assimilate technology imported from abroad. In the context of metal working, the technology which Indian machine tool builders imported during the 1950s and 1960s, provided the initial bulk to assimilate and start in-house R&D towards the advanced development of the products or to attempt the design and development of absolutely new products and processes. Organised production of standard machine tools commenced in the 1950s in collaboration with renowned foreign machine tool builders. In a bid to get the best in the world, collaborating firms from West Germany, Switzerland, UK, USA, France and Italy were selected to produce many general purpose machine tools under licence. By and large, even during sixties, much of what the Indian machine tool industry produced was based on imported know-how. But right in the early years of that decade some machine tool manufacturers started their own design and development sections because they were aware of the pitfalls of permanently depending on foreign source for the know-how to design, develop and produce machine tools. Around the same time, the Central Machine Tools Institute was set up in Bangalore to help the Indian machine tool industry in the field of R&D.

In the beginning, the Indian designers faced many hurdles, and progress was slow. The first indigenously designed centre lathe of this period and an electrically controlled milling machine took a long lead time to bring them to the level of production. But the important thing was that a beginning had been made, and the pride of achievement spurred the Indian designers to attempt more. This led to the indigenous design and development of a number of machine tools during the 1970s. It was at this time that many other manufacturers in the medium sector also started best owing attention on productionizing their own designs, or those obtained from the Central Machine Tool Institute in Bangalore.

Even as the Indian R&D engineers were striving hard to master the techniques of designing general purpose machine tools during the sixties, the developments abroad were aimed at exploring new frontiers. The advent of numerical control or NC, which was first tried on aircraft profilers in USA, had spread to a number of other applications like turning, milling and drilling. A new breed of machine tools like NC lathes, NC toolroom milling machines and NC machining centres made their debut, promising levels of productivity that were just inconceivable with general purpose machine tools. The advent of NC threw the manufacturing scene into a ferment. The astounding pace of developments in electronic hardware and software made the NC machine tools valuable production aids. Automation was pushed to higher levels achieve unfailing accuracy and higher productivity. Computer techniques led to a computer aided design (CAD) and computer-aided manufacturing (CAM). Robots were developed to relieve man from working in hazardous environments. Robots were integrated into the manufacturing chain to facilitate continuous production, boosting productivity to very high levels. In fact, the philosophy of development was to achieve higher productivity R&D in machine tools involved a multi-disciplinary approach a blend of mechanics and electronics. The essence of development was the shift from single purpose machine tools to combining several operations in one machine by providing the tools, changing them automatically and even automating job loading and unloading. Add to this the features of self-monitoring, diagnostics and fault correction, the end result is the advanced machine tool of today. This is the picture of developments from 1970 onwards.

Indian R&D engineers continued the task of designing and developing more advanced general purpose machine tools like copying lathes, precision, grinding machines, special purpose machines and gear cutting machines during the decade of 1970s. They did not remain indifferent to technological developments taking place abroad. In 1973 the first NC lathe was built in India, and this was followed by the first Turret Machining Centre in 1974. These machines were, however, interfaced with imported NC controls. By 1979 the technological strength of Indian R&D engineers was even more evident when they exhibited the first slant bed NC lathe and vertical machining centre in the IMTEX exhibitions. By 1982, development efforts had touched even sophisticated CNC machining centre with auto-tool changer. Turning centres and CNC vertical turning centres, all incorporating advanced features. But the country still relied heavily on imported controls. A beginning has now been made to built systems indigenously.

Although CNC technology was firmly established in advanced countries by mid-70s, its development and application in India was rather tardy. The global view of technological developments in machine tools is replete with the sophistication of CNC machines such as CNC turning centres, CNC grinding centres, gear making machines and CNC electro discharge machines. These CNC machines are the building block of future manufacturing systems. Flexibility in production is now sought to be introduced by linking these CNC machines with automatic work and tool handling systems such as robots and robotrailers, automatic inspection systems and fault diagnosis systems.

In India, the first CNC machine was introduced in the machine tools industry almost a decade ago. In the field two institutions have assumed a pioneering role. They are the Central Machine Tools Institute (CMTI) on research and design side, and the public-sector Hindustan Machine Tools (HMT) on manufacturing side. Both these institutions located in Bangalore have been contributing to the growth of Indian machine tools industry and introduction of latest technology in the field. The present range of indigenous CNC machines include turning centres, and grinding centres. Many other machines such as CNC gear cutting machine, CNC wire out EDM machines and CNC turrent punch presses are under various stages of development.

A number of machine tools including numerical controlled (NC) systems and machines along with necessary hardware and software, accessories, cutting tools and productivity aids have been designed and developed. Many of these machines are under regular production under licence from the institute. The NC division of CMTI has played a key role in introduction of NC machines in the country and is running an intensive development programme in introduction of CNC systems.

The institute is acting as a prime R&D organisation providing technical assistance to the machine tool and engineering industry in the areas of design development, research, manufacturing technology, numerical control, computer aids, testing standards, training and technical information. It has recently developed a decladding machine for profile machining of three dimensional curved surfaces to be used in atomic reactors. Horizontal band saw machine, profile dressing attachment, hydraulic copying equipment for machining cams for earth moving equipment and boring and facing head are some of the other design projects which the institute had completed recently.

The machine tools now under development of CMTI include a precision tool milling machine, a precision lathe, and a CNC cylindrical grinder. The institute has also successfully carried out an interesting project on image processing for object identification using a television camera. This experiment has potential application for development of robotics. The institute is also working on development of a robot for industrial applications.

During the past five years, at least three organisations have initiated development of CNC technology within the country. The CNC systems have been interfaced with machine tools built by leading machine builders. Thus the technological base for CNC systems is now available in the country. But the commercial production capability for CNC machines is yet to be established fully. With an anticipated requirement of about 100 systems per year, the Indian industry needs to make more determined efforts to build indigenously the CNC systems. Since CNC systems constitute a rapidly advanced high technology product the base of development has to be speeded up.

As a result of research and development efforts of CMTI and HMT, the Indian machine tool industry has a reasonably strong technological base. Numerically controlled machines were developed in the country as back as in 1972, and the basic features and specification of the indigenously developed CNC machines since then have been competitive compared to their foreign counterparts. But the emerging market of CNC machines in the country demands more sophistication in operation, controls, precision and flexibility.

Besides CNC machines, another equally exciting development has been the appearance of robots on the metal working scene in a big way during the past five years. Robots have been seen as an answer to the productivity challenge in metal working operations and are already in use on the shop floor for machine tending (component loading and unloading). This trend would further strengthen with coming of unmanned systems and Flexible Manufacturing Systems (FMS), where efficiency of component loading and unloading would be an important criterion. The current trend in robots has been so strong that these developments might become practicable much earlier than expected.

In India, a few research groups have been working on development of robots, but a breakthrough is yet to be made in the field of robots for large scale industrial application. The groups working on robotics include R&D of HMT, CMTI, the Indian Institute of Technology (IIT), Madras, Indian Institute of Science (IIS), Bangalore, and the Hyderabad Science Society, Hyderabad. A few of the engineering colleges have also been working on development of robot arms and sensors.

The public sector defence production unit, Bharat Electronics Limited (BEL), was the first Indian industry to introduce an indigenously developed robot in its production line. It introduced a "pick and place" type robot with a three axis movement in its second television tube production plant at Bangalore. The pneumatically operated robot was incorporated to improve the productivity while maintaining quality in the semi-automated plant, climinating operator fatigue and other health hazards. Controlled by a microprocessor the robot collects coated TV bulbs from two continuously rotating screen coating equipment and unload them on to an indexing type twelve-headed drying machine. It is also capable of correcting any possible errors due to difference in speeds of the two coating equipment to get overall synchronised movements between the three machines. The display board on the control panel indicates the exact functional area where the fault has occurred to facilitate instant trouble shooting.

The robot control release of coated bulbs from vacuum chucks of the coating machines rotates the lower arms to predetermined positions, centres the neck of the TV bulb on drying equipment. The presence of absence of a bulb in any of the 12 positions is sensed by light modulated photosensors, which act positively, irrespective of the intensity of the ambient light. Magnetic sensors sense the desired positions of various electro-pneumatic end-effectors. Provisions have also been made in the central control unit of the robot to add more functions down the line if required. Microprocessor controlled scanners scan temperatures in various zones of the screen back and the in-line exhaust system and flash back temperature indications.

The Hyderabad Science Society, which has done pioneering work in the field of robotics, developed the country's first robot few years back. It has also developed a robot with the ability to move about using ultrasonic sensors just like a bat. The ultrasonic sensor developed by the society team, has been attached to walrobo-two, a robot developed by the society earlier. The robot with ultrasonic sensors could be used in industries involving health hazards such as spraying, chemical and nuclear plants. Tactile sensors to help robots pick up objects and infrared sensors for sensing profile have also been developed. All these extra facilities will be incorporated finally into robotic manipulator.

The robotics laboratory at the school of automation of the Indian Institute of Science (IIS), Bangalore, has developed a microprocessor-based robotic arm. The arm has four joints, each of which is actuated by a stepper motor. The robotic arm can be controlled manually from a front panel or automatically through a video terminal and a keyboard. The microprocessor has been programmed in such a way that a lay man can control the robotic arm. It can be commanded to pick an object from a specified location or to search, locate and then grip an object within the work area.

All these developments in the newer and exciting areas of CNC machines and robotics are likely to take concrete shape in future and might help in increasing productivity and safety in various industries in the country.

An exercise in technology forecasting conducted in India with special reference to the machine tool industry brought out scenarios and projected the gaps to be bridged. A recent exercise in perspective planning has resulted in a number of recommendations to improve R&D base to make the machine tool industry self-reliant by the end of this decade.

Given the will to pursue R&D work in the spheres in which we are lacking, it will certainly make the industry self-reliant. A healthy and self-supporting machine tool industry will make the metal-working industry and other user industries, like automobile, machine-building, electrical and other engineering industries, more dynamic leading to a tremendous industrial resurgence. That happened in USA, West Germany and Japan and can happen in India also, given the national commitment to foster R&D in all areas of vital importance.

BIBLIOGRAPHY

1. National Committee on Science and Technology, *Summary of the Draft Plan of the NCST Group on Education and Scientific Research,* New Delhi, 1973, p. 19.

2. National Committee on Science and Technology, *Science and Technology Plan, 1974-79,* Vol. I and Vol. II (Draft) Government of India, New Delhi, 1973, p. 517.

3. National Committee on Science and Technology, *An Approach to the Science and Technology Plan (Background Documents),* Meeting Plan Frame for Science and Technology (1980-85), New Delhi 1980, p. 250.

4. Government of India, *Report of the Working Group on Science and Technology for the Eighth Plan, 1990-95,* Planning Commission, New Delhi, 1995, p. 82.

5. Government of India, *Technology Policy Statement,* Department of Science and Technology, New Delhi, 1983, p. 16.

6. Government of India, *Science and Technology in India,* Department of Science and Technology, New Delhi, 1983, p. 48.

7. Government of India, *Promotion and Support to Indigenous Technology,* Department of Science and Technology, New Delhi, 1983, p. 73.

8. Government of India, *Incentives for Scientific Research and Indigenous Technology,* Guidelines, Department of Science and Technology, New Delhi, 1983, p. 22.

9. Government of India, *Thrust Areas in Engineering Science,* Department of Science and Technology, New Delhi, 1983, p. 59.

10. Government of India, *Promotion and Development of Technology for Scheduled Castes/Scheduled Tribes and other Weaker Sections* — Promotional Scheme : General Information and Direction & Approach, Department of Science and Technology, 1984, New Delhi, p. 12.

11. Government of India, *Science and Technology for Women* (A revised illustrated version of the compendium of technology), Centre of Science for Villages, Wardha, Department of Science and Technology, New Delhi, 1983, p. 169.

12. Government of India, Regional Sophisticated Instrument Centres: General Information, Department of Science and Technology, New Delhi, 1982, p. 17.

13. Government of India, Projects on Environmental Research, Department of Environment, New Delhi, p. 31.

14. Government of India, A National Policy on Environment, Department of Environment (B.P. Pal), New Delhi, 1981, p. 15.

15. Government of India, *Air Quality, Management and Energy Policies,* Department of Environment (N.L. Ramanathan, J.M. Dave and T.A. Siddique), New Delhi, 1981, p. 48.

16. Government of India, *R&D Institutions in India : A Directory of Addresses,* Department of Science and Technology, New Delhi, p. 160.

17. Government of India, *Directory of In-House R&D Units* Recognised by Deparatment of Science and Technology, Department of Technology, New Delhi, 1982, p. 90.

18. Ministry of Science and Technology, *Directory of Recognised In-House R&D Units,* Department of Scientific and Industrial Research, New Delhi, 1985, p. 95.

19. Government of India, Report of the Committee for Recommending Legislative Measures and Administrative Machinery for Ensuring Environmental Protection, Department of Science and Technology, 1980, p. 76.

20. Government of India, National Committee on Environ-Planning and Coordination : Recommendations and Activities, Department of Environment, New Delhi, 1982, p. 23.

21. Ministry of Labour, *Occupational and Educational Pattern in India, 1979,* Directorate General of Employment and Training, New Delhi, 1985, p. 122.

22. Ministry of Labour and Rehabilitation, *Occupational and Educational Pattern in India* : Public Sector : 1978, Directorate General of Employment and Training, New Delhi, 1984, p. 136.

23. Indian Council of Agricultural Research, *Research Highlights, 1983,* Publications and Information Division, New Delhi, 1984, p. 64.

24. Government of India, Housing and Construction Technology: *Draft Status Report,* National Committee on Science and Technology, New Delhi, 1975, p. 880.

25. Government of India, Report of the Working Group on Energy Policy, Planning Commission, New Delhi, 1979, p. 121.

26. Central Electricity Authority, Report of the Sub-Group on Science and Technology in Power Sector, Planning Wing, New Delhi, 1984, p. 613. (Appendices).

27. Government of India, Committee on Transport Policy and Coordination : Final Report, Planning Commission, New Delhi, 1966, p. 340.

28. Government of India, Report of the National Transport Policy Committee, Planning Commission, New Delhi, 1980, p. 393.

29. Government of India, Annual Report, 1982 83, Department of Science and Technology, New Delhi, p. 176.

30. Ministry of Science and Technology, Annual Report, 1984-85, Department of Scientific and Industrial Research, New Delhi, 1985, p. 88.

31. National Research Development Corporation, 27th Annual Report, 1980-81, New Delhi, p. 82.

32. Government of India, *Sixth Annual Report of the Comptroller General of Patents, Designs and Trade Marks*, 1977-78, Calcutta, 1978, p. 41.

33. Council of Scientific and Industrial Research, *Annual Report, 1983-84*, National Institute of Science, Technology and Development Studies, New Delhi, 1984, p. 48.

34. Government of India, *Research and Development Statistics, 1978-79*, Department of Science and Technology, New Delhi, 1979, p. 162.

35. Government of India, *Research and Development Statistics, 1980-81*, Department of Science and Technology, New Delhi, 1981, p. 181.

36. Government of India, *Research and Development in Industry, 1980-81*, Department of Sceince and Technology, New Delhi, 1981, p. 37.

37. Government of India, *Research and Development in Industry, 1982-83*, Department of Science and Technology, New Delhi, 1983, p. 55.

38. Ministry of Agriculture, *Annual Report 1983-84*, Department of Agricultural, Research and Education, New Delhi, 1984, p. 114.

39. Council of Scientific and Industrial Research, *CBRI Annual Report, 1983*, Central Building Research Institute, Roorkee, 1983, p. 80.

40. Government of India, *Report 1983-84*, Ministry of Works and Housing, New Delhi, 1984, p. 59.

41. Government of India, National Paper: India UN Conference on New and Renewable Source of Energy (Nairobi, 1981), Department of Science and Technology, New Delhi, 1981, p. 32.

42. Government of India, India the Nairobi Programme (The Development and Utilisation of New and Renewable Source of Energy in India), Department of Science and Technology, New Delhi, 1982, p. 18.

43. Government of India, *Renewable Energy in Action*, Commission for Additional Sources of Energy, New Delhi, p. 23.

44. Government of India, *Report 1983-84*, Department of Petroleum, New Delhi, 1984, p. 39.

45. Ministry of Irrigation and Power, *Report 1984-85*, Department of Power, New Delhi, 1984, p. 39.

46. Ministry of Energy, *Annual Report 1983-84*, Department of Non-Conventional Energy Sources, New Delhi, 1984, p. 58.

47. Central Board of Irrigation and Power, *Annual Review 1984*, (Summaries : Research Scheme on Power), New Delhi, 1983, p. 45.

48. Ministry of Industry, *Report 1983-84*, Department of Industrial Development, New Delhi, 1984, p. 91.

49. Ministry of Industry, *Report 1982-83*, Department of Industrial Development, New Delhi, 1983, p. 184.

50. Ministry of Industry, *Report 1981-82*, New Delhi, p. 241.

51. Government of India, *Annual Report 1984-85*, Ministry of Shipping and Transport, New Delhi, 1984, p. 76.

52. Government of India, *Annual Report 1983-84*, Ministry of Shipping and Transport, New Delhi, 1984, p. 76.

53. Karol I. Pelc, Science/Technology Policy and Forecasting, Centre for the Study of Science, Technology and Development (CSIR), New Delhi, 1979, p. 16.

54. Nandini Joshi, *Manpower Planning in Scientific Research Institute*, Centre for Study of Science, Technology and Development (CSIR), New Delhi, 1979, p. 76.

55. S. Pruthi and P.S. Nagpaul, *Pattern and Role of Informal Communication in R&D*, Centre for Study of Science, Technology and Development (CSIR), New Delhi, 1978, p. 49.

56. M.S. Randhawa, *Natural Resources of India*, Planning Commission, New Delhi, 1963, p. 24.

57. V. Rao, Aiyagiri and P.L. Lavakare, *Policy Formulation in Science and Technology*, Department of Science and Technology, New Delhi, 1981, p. 48.

58. A Rahman. *The Cultural and Philosophical Roots of Scientific Tradition,* National Institute of Science, Technology and Development Studies (CSIR), New Delhi, 1983, p. 25.

59. A Rahman (ed.) *Science and Technology in Indian Culture — A Historical Perspective,* National Institute of Science, Technology and Development Studies (CSIR), New Delhi, 1984, p. 251.

60. A. Rahman, *Science and Technology in India,* National Institute of Science, Technology and Development Studies (CSIR), New Delhi, 1984, p. 212.

61. A. Rahman and N. Haritash. *The Role of Parliament in the Formulation of National Science and Technology Policy,* National Institute of Science and Technology Policy, National Institute of Science, Technology and Development Studies (CSIR), New Delhi, p. 108.

62. Acharya, P.K., *An Encyclopedia of Hindu Architecture* (Oxford, 1946).

63. Basham, A.L., *The Wonder That was India* (New York, 1954).

64. Brown, Percy, *Indian Architecture* (Buddhist and Hindu Periods), (Bombay, 1942).

65. *Cambridge History of India, ed.* by Edward J. Rapson (Cambridge 1922).

66. Central Road Research Institute, *History of Road Development in India* (New Delhi, 1964).

67. Chakravarty, Toponath, *Food and Drink in Ancient Bengal* (Calcutta, 1959).

68. Garratt, Geoferey T. (ed.) *The Legacy of India,* (Oxford, 1937).

69. Gupta H.C. Das, "Bibliography of Prehistoric Indian Antiquities", *Journal of the Asiatic Society of Bengal,* n.s. 27, (1931), 5-96.

70. Lal, B.B., "Further Copper Hoards from the Gangetic Basin and a Review of the Problem", *Ancient India* 7(1951), 20-30.

71. Macgeorge, G.W. Ways and Works in India, Being an Account of the Public Works in that Country from the Earliest Times up to the Present Day (London, 1894).

72. Majumdar, R.C., "Scientific Spirit in Ancient India" in the *Evolution of Science,* ed. by Guy S. Metraux and Francois Crouzet (New York 1963); reprinted from Journal of World History, VI 2.

73. Marshall, John H., *Mohenjo-daro and the Indus Civilization;* being an official account of archaeological at Mohenjo-daro carried out by the Government of India between the years 1922 and 1927 (3 Vols., London, 1931).

74. Piggot, Stuart, *Prehistoric India* (Harmondsworth, England, 1950).

75. Ray, P.C., *History of Chemistry in Ancient and Medieval India* (new ed. Calcutta, 1956).

76. Subbarao, Bendapudi, *The Personality of India* (Baroda 1956).

77. Swarup, Shanti, *The Arts and Crafts of India and Pakistan* (Bombay 1957).

78. Toy, Sidney, *The Fortified Cities of India* (London 1965).

79. Wheeler Mortimer, *Civilizations of Indus Valley and Beyond* (London 1966).

80. White Lynn, Jr., "Tibet, India and Malaya as Source of Western Medieval Technology, *American Historical Review,* 45 (1960), p. 515-526.

81. Ota Sulc, *Methodology of Forecasting Complex Development Processes of the Scientific and Technological Revolution,* Centre for the Study of Science, Technology and Development (CSIR), New Delhi, 1977, p. 46.

82. M.S. Patwardhan, "Chambers of Commerce in 1980s", *Commerce,* November 21, 1981, pp. 927-929.

83. P.S. Shankar and R.S. Sharma, "The Technology Policy in Relation to the Industrial and Science Policies in India", *Journal of Scientific and Industrial Research,* Vol. 42, February 1983, pp. 51-63.

84. Technology Forecasting Mechanism: TPIC Submits Blueprint, *The Economic Times,* September 7, 1985.

85. M. Dayal, *Energy Policies and the Contributions of Renewable* Energies in India, Indian Association for the Advancement of Science, New Delhi, p. 32.

86. Commissioner of Labour in Karnataka, *Karnataka Labour Journal* Vol. XIX No. 1-4, January-February-March 1985, p. 188.

87. Special issue on Energy, *Journal of the Institution of Engineers* (India), Vol. 33, No. 6, December 1983.

88. *Nature,* Vol. XV, April 12, 1984.

89. Nehru and Science, *Interdisciplinary Science Period,* Vol. 10, No. 2, 1985, pages 105-113.

90. Mohan Bhatia, *S&T Manpower Planning,* Centre for Studies in Science Policy, Jawaharlal Nehru University, New Delhi, 1982, p. 9.

91. Mohan Bhatia, *System Approach to Science and Technology Plan,* Centre for Studies in Science Policy, Jawaharlal Nehru University, New Delhi, 1976, p. 7.

92. B.V. Ranga Rao, *Science Policy for Development,* Centre for Studies in Science Policy, Jawaharlal Nehru University, New Delhi, 1976, p. 443.

93. B.V. Ranga Rao, *Evolution of Apex Science Policy Body in India,* Centre for Studies in Science Policy, Jawaharlal Nehru University, New Delhi, p. 16.

94. Satish C. Seth, *India Towards 21st Century* (Unpublished).

95. S. Varadarajan and K.V. Swaminathan, *Technology Forecasting* (A Paper for Consideration of TPIC, Unpublished), 1984.

96. National Institute of Science, *Technology and Development Studies,* International Seminar on the Role of Fundamental Research in Developing Countries : Report and Recommendations, CSIR, New Delhi, 1983, p. 15.

97. National Institute of Science, Technology and Development Studies, Seminar of Frontier Technologies in the Third World, Towards a Strategy of Self-Reliance : A Repprt. CSIR, New Delhi 1982, p. 23.

98. International Development Research Centre, *The Science and Technology Policy Instruments Project*, Ottawa.

99. *Science in India,* Department of Science and Technology.

100. *Glimpses of Indian Technology*, Bal Krishna Publications Division, 1988.